Student Manual to Accompany

EMERGENCY MEDICAL RESPONSE TO HAZARDOUS MATERIALS INCIDENTS

Richard H. Stilp

Armando S. Bevelacqua

Delmar Publishers

an International Thomson Publishing Company I(T)P®

Albany • Bonn • Boston • Cincinnati • Detroit • London • Madrid
Melbourne • Mexico City • New York • Pacific Grove • Paris • San Francisco
Singapore • Tokyo • Toronto • Washington

NOTICE TO THE READER

Publisher does not warrant or guarantee any of the products described herein or perform any independent analysis in connection with any of the product information contained herein. Publisher does not assume, and expressly disclaims, any obligation to obtain and include information other than that provided to it by the manufacturer.

The reader is expressly warned to consider and adopt all safety precautions that might be indicated by the activities described herein and to avoid all potential hazards. By following the instructions contained herein, the reader willingly assumes all risks in connection with such instructions.

The publisher makes no representations or warranties of any kind, including but not limited to, the warranties of fitness for particular purpose or merchantability, nor are any such representations implied with respect to the material set forth herein, and the publisher takes no responsibility with respect to such material. The publisher shall not be liable for any special, consequential or exemplary damages resulting, in whole or in part, from the readers' use of, or reliance upon, this material.

Delmar Staff

Publisher: Robert Lynch
Acquisitions Editor: Mark Huth
Developmental Editor: Jeanne Mesick
Production Coordinator: Toni Bolognino
Art and Design Coordinator: Michael Prinzo

Printed in the United States of America
For more information, contact:

Online Services

Delmar Online
For the latest information on Delmar Publishers new series of Fire, Rescue and Emergency Response products, point your browser to:
http://www.firesci.com

A service of I(T)P®

Delmar Publishers
3 Columbia Circle, Box 15015
Albany, New York 12212-5015

International Thomson Publishing Europe
Berkshire House 168-173
High Holborn
London, WC1V7AA
England

Thomas Nelson Australia
102 Dodds Street
South Melbourne, 3205
Victoria, Australia

Nelson Canada
1120 Birchmount Road
Scarborough, Ontario
Canada M1K 5G4

International Thomson Editores
Campos Eliseos 385, Piso 7
Col Polanco
11560 Mexico D F Mexico

International Thomson Publishing Gmbh
Königswinterer Strasse 418
53227 Bonn
Germany

International Thomson Publishing Asia
221 Henderson Road #05-10
Henderson Building
Singapore 0315

International Thomson Publishing - Japan
Hirakawacho Kyowa Building, 3F
2-2-1 Hirakawacho
Chiyoda-ku, 102 Tokyo
Japan

1 2 3 4 5 6 7 8 9 10 XXX 02 01 00 99 98 97

Library of Congress Cataloging-in-Publication Data 96-34053

ISBN #0-8273-7830-0

CONTENTS

PREFACE

This manual is designed to follow the textbook *Emergency Medical Response to Hazardous Materials Incidents*. While working through this book, evaluate each answer and choose the most appropriate one. It is designed to follow the text in the presented order so that you can find the answer when confronted with a difficult task. The material presented in the text is thoroughly reviewed within this guide. In order to gain the most from the material, the authors have included a step by step approach in its presentation.

This manual was designed to take you through a progression of information starting with the basic principles of hazardous materials response where basic awareness and operations level training are presented, and reinforcing these basic principles through the use of questions, activities, and scenario work.

A chapter was included examining the use of an incident management model and how this model is utilized on emergency incidents, particularly on hazardous materials alarms. An example of standard operating guidelines for hazardous materials response is available from the publisher through the World Wide Web. Updates to this SOG will also be available on WWW. This is provided so that the team may have a guideline for producing its own management system utilizing the points covered in this chapter.

Section two gives you a detailed didactic overview of chemistry and toxicology, the principles of which are interrelated and complex. It is intended to aid the emergency worker toward *enemy identification*, i.e. the hazardous material, and is directed toward referencing of the material, medical applications, and the information required for overall scene management. Each discipline is approached as an isolated field with a section in this manual on how chemistry and toxicology can be applied to the incident.

Once you have studied the basic principles, chemistry, and toxicology, section three will introduce to you the principles of treatment. This section identifies the anatomy and physiology as it relates to hazardous materials exposures. The chapter then moves on to some of the more common poisonings and the treatment for these hazardous materials events. The student manual will identify possible avenues of procedure, understanding that there are no absolutes in treating exposed patients. The manual allows you the freedom to explore treatment of exposed patients and also allows you the freedom to examine alternate avenues of assessment and treatment, keeping in mind the principles and the recognized applications that are presented.

Section four identifies the protection controls you may incorporate within your respective system. These, like previous chapters, have example standard operating guidelines written for the emergency provider's use. The student manual, through the use of activities, will demonstrate the intended applications.

Section five of the textbook incorporates those areas that are of special interest but may not be needed within your system. These are ancillary considerations of emergency response. Much of the information presented in the previous chapters will become incorporated within the activities and scenario work included in this section. It is recommended that you, as a student, start with chapter one and complete the text and manual in the order given. Even if you have had formal hazardous materials training in the past, basic concepts should be reviewed. It would benefit you the most if you organize your work according to the following suggestions:

1. Read the chapter that correlates with the student manual.
   - **A.** When reading the textbook, be able to identify the objectives that are found in the beginning of each chapter.
   - **B.** Take note of safety issues, caution concerns, and special notes.
2. After reading the corresponding chapter, read and complete the scenario found at the end of each chapter.
3. Once you have completed the scenario and identified the objectives, turn to the chapter found within this manual and complete, by circling the appropriate answer, or writing your responses on the write-on lines.
4. Now go back to the textbook and find the answers, comparing your responses. If your response is different, analyze why you chose that particular answer. In some cases, this may require you to reread that particular section in order to have a firm understanding of the material.

A certain level of comprehension of terms and principles will be essential in order for you to become functional in this area at the hazardous materials incident. While reading this material, memorize terms and abbreviations along with the corresponding principles. Answers to the questions are found within the instructor's guide and should be discussed with the delivery instructor.

The entire book, in combination with this student manual and a delivered class, is designed to give you the nuts and bolts of hazardous materials emergency medical response. Most of the text can be delivered to both Emergency Medical Technicians (EMTs) and paramedical level response personnel. The chapter on treatment will require education on the paramedical level. However, all will benefit from the information presented. None of the information is beyond the understanding of the emergency response worker. In jurisdictions that require advanced training within this discipline, all who attend will gain insight needed for these types of responses.

Hazardous Materials Concepts

Objectives

Given a hazardous materials incident, you should be able to describe the general roles, responsibilities, and hazards presented to the responder and victims, appropriately describing the importance of a planned response.

As a hazardous materials medical technician, you should be able to:

- Identify the responsibilities and roles of the Local Emergency Planning Committee (LEPC) as they relate to the hazardous materials event.
- Describe the functions of the emergency dispatch center.
- Define size-up as it relates to the hazardous materials incident.
- Identify basic referencing techniques, discussing the pros and cons of these referencing techniques.
- Describe the DOT placard, identifying its strengths and weaknesses.
- Identify and define NFPA 704.
- Describe the four identifiers of the placard and the DOT hazard classes.
- Identify the zones and describe the functions within each zone.
- Describe the three components necessary to create a plan.
- Identify and define the three levels of a hazardous materials incident.
- List the internal and external elements of resource management.
- Define the standard of care as it relates to a hazardous materials incident.
- List the four elements of negligence and define each element.

CONTENT QUESTIONS

1. Hazardous material is defined as a hazardous substance or material that may be potentially harmful to the public's health or welfare if it is discharged into the environment. Which department does this definition fall under?
   - **A.** Department of Transportation (DOT)
   - **B.** Environmental Protection Agency (EPA)
   - **C.** Occupational Safety and Health Administration (OSHA)
   - **D.** American Conference of Governmental Industrial Hygienists (ACGIH)

2. In an act dated back to the mid seventies, which department defines hazardous material as any substance or material in any form or quantity that poses an unreasonable risk to safety and health and to property when transported in commerce?
   - **A.** Department of Transportation (DOT)
   - **B.** Environmental Protection Agency (EPA)
   - **C.** Occupational Safety and Health Administration (OSHA)
   - **D.** American Conference of Governmental Industrial Hygienists (ACGIH)

3. Which department views a hazardous material from a potentially hazardous standpoint? They rate conditions that may cause injury or death, whether they are obvious or not, as they are found in the working environment.
   - **A.** Department of Transportation (DOT)
   - **B.** Environmental Protection Agency (EPA)
   - **C.** Occupational Safety and Health Administration (OSHA)
   - **D.** American Conference of Governmental Industrial Hygienists (ACGIH)

4. Considerations that should always be in the minds of the emergency responder are
   - **A.** personal safety.
   - **B.** identification and evacuation.
   - **C.** secondary exposure.
   - **D.** all the above.

5. The basic step before any objective or task can be accomplished is called
   - **A.** management of the incident.
   - **B.** information about the incident.
   - **C.** preplanning the incident.
   - **D.** controlling the risk.

6. In the general sense, there are several basic components that create the preplanning scheme. Each of these components has areas of overlap that can be thought of as a loop of planning activity. These concepts are:
   - **A.** 2,4,5,3.
   - **B.** 1,5,2.
   - **C.** 2,4,1,3.
   - **D.** 1,5,3.

   1. The control of future risks
   2. Preplanning
   3. Adequate range of information before the event
   4. Response levels
   5. Management of available resources

7. There are several components that make up controlling the risk; these are:

A. 1,2,3,4,5,6.	1. Identification of the risks
B. 6,4,2,3,1.	2. Developing a workable plan
C. 6,1,4,3,5.	3. Available resources
D. 1,5,2,4,6.	4. Implementation of the plan
	5. Identify public and private entities
	6. Annual review

8. Which of the following examines major roadways, railroads, Tier II reports, and general industrial processes within the community?
   A. Identification of public entities
   B. Development of resources
   C. Identification of the risks
   D. Hazard categories

9. There are risk levels within any community. Match the risk assessment with the appropriate hazard type.

A. Common household items	1. Hazard Type A
B. School laboratories	2. Hazard Type B
C. Transportation accidents	3. Hazard Type C
D. Occupancies with potential quantities	4. Hazard Type D

10. A good place to start when assessing the community for hazard types is utilizing which of the following?
    A. Material Safety Data Sheets
    B. Placards and labels
    C. Tier II reports
    D. A and C

11. In the simplest sense, the evaluation of life, environment, and property are priorities, however this evaluation is the analysis of life safety and the environmental issues that surround this segment.
    A. Risk assessment
    B. Hazard analysis
    C. Statistical analysis
    D. Vulnerability

12. The outlined approach toward what could happen "if", the process of evaluating plausible scenarios based on known facts and information, is called
    A. risk assessment.
    B. hazard analysis.
    C. statistical analysis.
    D. vulnerability.

13. The level of an incident identifies the resources available within the planning stages. When an incident occurs that the first responding agency or local system can handle, it is called a
    A. Level I incident.
    B. Level II incident.
    C. Level III incident.
    D. Level IV incident.

14. The level of an incident identifies the resources available within the planning stages. When the incident has become potentially hazardous, involving a large population segment, the condition is either a public health hazard, or has the potential to become a danger to life and the environment, it is called a
    A. Level I incident.
    B. Level II incident.
    C. Level III incident.
    D. Level IV incident.

15. The level of an incident identifies the resources available within the planning stages. When the situation has the high probability of becoming a serious health hazard to human life and/or will affect the environment, it is called a
    A. Level I incident.
    B. Level II incident.
    C. Level III incident.
    D. Level IV incident.

16. One must identify the public and private entities that may be needed. As the incident progresses, the need to utilize outside resources will increase. Outside resources are usually thought of as private contractors, however, within municipalities and county government there exist individuals who can also assist at hazardous materials events. These are termed inside and outside resources. Under what preplanning activity would this be placed?
    A. Identification of public entities
    B. Development of resources
    C. Identification of the risks
    D. Hazard categories

17. Developing a workable plan encompasses what types of goals?
    A. Mission statements
    B. Short term goals and long term goals
    C. The local agencies' ability to serve and protect, delivery of service
    D. Personnel accountability

18. The receiving hospital can handle any chemically exposed patient that may be presented if preplanned properly.
    A. True
    B. False

19. When true referencing is to be done at the scene of a hazardous materials event, how many reference sources should be consulted?
    A. 2 B. 3 C. 1 D. 4

20. In order to have a workable plan, what should be utilized during and after implementation of the plan?
    A. Involvement of other departments
    B. Annual revision
    C. Input from field operation personnel
    D. All the above

21. What are the elements of resource management?
    A. Internal elements
    B. External elements
    C. A only
    D. A and B

22. What internal elements should be considered when discussing the plan?
    A. Building department, mutual aid agreements, police department, schools
    B. Building department, mutual aid agreements, EMS agencies, police departments
    C. Building department, mutual aid agreements, police department, water and sewer department
    D. Building department, mutual aid agreements, police department, hospitals

23. For the Level II and III type incidents, surrounding emergency response groups must be placed within the plan. What is this segment of the plan called?
    A. Contracting for service
    B. Mutual aid agreements
    C. Resource management
    D. Automatic response

24. The law enforcement agencies will be needed to provide what function?
    A. Block off areas of involvement
    B. Aid in evacuation routes
    C. Maintain the orderly conduct of the community as a whole
    D. All of the above

25. Schools can be utilized effectively for a shelter except under what condition?
    A. When the facility does not provide the space required for the evacuees
    B. When the facility is within the contaminated area
    C. When the plan has not identified such a facility
    D. All the above

26. What are the external elements of resource management?
    A. Contractors, hospitals, EMS agencies, chemical engineers and toxicologists, fire brigades
    B. Contractors, hospitals, fire agencies, chemical engineers and toxicologists, fire brigades
    C. Contractors, hospitals, EMS agencies, chemical engineers, fire brigades
    D. Contractors, hospitals, fire agencies, chemical engineers, fire brigades

27. What external elements are at a high risk for secondary exposure?
    A. Contractors
    B. Hospitals
    C. EMS agencies
    D. B and C

28. What are the data points under data analysis?
    A. Demographics, chemical, occupancy, pre-fire plans, financial, legislative
    B. geographic, toxicological, occupancy, pre-fire plans, financial, legislative
    C. Demographics/geographic, chemical/toxicological, occupancy, pre-fire plans, financial, legislative
    D. Demographics/geographic, chemical/toxicological, occupancy, pre-fire plans, legislative

29. Occupational Safety and Health Administration standard is called
    A. 1910.200.
    B. 1910.121.
    C. 1910.120.
    D. 1920.110.

30. The OSHA regulation provided what general aspect of hazardous materials?
    A. A regulation that describes placarding, MSDS, and labels
    B. A regulation that defines the operational goals of emergency responders
    C. A regulation that governs exposures to employees
    D. A regulation that details the announcement of hazardous materials

31. What is the difference between a law and a regulation?
    A. A regulation is enacted by congress and signed by the president
    B. A law is enacted by congress and signed by the president
    C. A regulation provides law
    D. A law is a standard

32. A law is enacted by congress and signed by the president. Regulations are those standards that are propagated by law, are not truly law but carry the same weight of law.
    A. True
    B. False

33. EMTs and paramedics do not have to follow any standard of care when dealing with individuals that have been exposed to a hazardous material.
    A. True
    B. False

34. Standards are procedures that are supported by which two general areas?
    A. Internal and external standards
    B. National Fire Protection Agency
    C. Occupational Safety and Health Administration
    D. Documentation standards

35. The 29 CFR 1910.120 and 40 CFR 311 are equal in terms of general regulations.
    A. True
    B. False

36. Size-up is a function that occurs
    A. upon arrival of a hazardous materials incident.
    B. upon arrival of an incident and every few minutes thereafter.
    C. initially upon arrival only.
    D. as a continuous process which starts within the dispatch center and ends after recovery.

37. What is the single most important function at the scene of a hazardous materials event?
    A. Isolation
    B. Identification
    C. Evacuation
    D. All are equally important

38. Concerning placards, which is a true statement?
    A. There should be a placard on both sides of the transporting vehicle.
    B. Placards are required on ALL hazardous materials carriers.
    C. Certain classifications require that the commodity is always placarded.
    D. There are three components to the placarding system.

39. The DOT placarding system has general components that will assist the responder in identifying the hazard classification. They are:
    A. Colored background, with centered numbers and compatibility codes
    B. Colored background, with centered numbers, compatibility codes, symbol at the bottom
    C. Colored background, with centered numbers, compatibility codes, symbol at the top, and sub-classification numbers

40. Labels and placards can be utilized interchangeably.

A. True **B.** False

41. The placard "Dangerous" is utilized under what conditions?

A. When the load is mixed and is less than 1001 pounds.
B. When the load is unknown and is less than 1001 pounds.
C. When the load is mixed and greater than 1001 pounds.
D. When the load is unknown and is greater than 1001 pounds.

42. Which of the following statements are true regarding the DOT placarding system?

A. The UN number can be utilized to identify the exact chemical.
B. The UN number will identify chemicals or chemical groups.
C. DOT placards are required for all transportation of hazardous materials.
D. The labels that look like placards must also be visible on all four sides of the transporting vehicle.

43. In the NFPA 704 placarding system, the blue stands for what hazard?

A. Toxicity **C.** Health
B. Medical **D.** Etiologic Agents

44. When dealing with a patient who has been exposed, what procedure considerations are there?

A. The transporting unit should have a plastic sheet covering the interior of the ambulance, all equipment covered, and rescuers in protective gear.
B. The transporting unit should have a plastic sheet covering the interior of the ambulance, all equipment covered, patient decontaminated, and rescuers in protective gear.
C. The transporting unit should have a plastic sheet covering the interior of the ambulance, all needed equipment covered, with removal of non-utilized equipment, and rescuers in protective gear.
D. The transporting unit should have a plastic sheet covering the interior of the ambulance, all needed equipment covered, with removal of non-utilized equipment, patient decontaminated, and rescuers in protective gear.

45. Helicopters have limited, if any, use at the hazardous materials event when dealing with patient care.

A. True **B.** False

CONTENT APPLICATION

You have been dispatched to a facility that processes large quantities of LPG. During the initial communication, the dispatcher states that several individuals are complaining of having difficulty breathing. The emergency operations center has now classified this response as a hazardous materials incident. Your arrival is moments away. You hear over the main dispatch frequency a full hazardous materials response complement. However, you know it will be several minutes before help arrives.

Your partner states that she will handle the isolation and initial evacuation parameters as set up in the prehazard planning book. She assigns you to the initial referencing of the chemical in question.

Upon arrival you notice a tractor trailer tanker issuing a vapor. The placard looks like 1670 or 1671, but from your distance it is very difficult to distinguish. Based upon this information, answer the following questions.

1. What are the chemical possibilities?

____

____

____

____

____

2. Which guide number will be utilized?

____

____

____

____

____

____

3. This event occurred minutes prior to your arrival. What are your isolation and evacuation areas for the above chemicals?

____

____

____

____

____

____

4. You have done all that is possible from a first response level, however you decide to reference these chemicals further assisting the hazmat team's science sector. What are the synonyms and trade names?

____

____

____

____

____

____

5. Your partner has been reassigned to medical sector during the initial stages of this incident. She asks you over the radio what health hazards she should expect from the patients. What are the signs and symptoms of the above placarded chemicals?

When dealing with a hazardous materials incident, preplanning is a vital component in establishing a community assessment. Answer the following questions utilizing your response department and surrounding resources.

1. Utilizing the information presented in this chapter, describe how you could apply the three general categories of the planning activity.

2. Identify the available resources that you would need during a level one, two, and three hazardous materials event.

3. Describe the data points that are needed during your planning of the hazardous materials response.

4. How would you apply the information learned during research of the above question in terms of short and long range goals?

____

____

____

____

____

____

____

____

5. Discuss the considerations that your agency should employ for a well planned response.

____

____

____

____

____

____

____

____

CHAPTER 1 SCENARIO

It is 5:00 P.M. on Saturday and you have just received a call about a suspicious odor in the warehouse area of your response district. The alarm was phoned in by a security guard working in a food distribution warehouse. She stated that the odor was unpleasant and caused an irritation in her throat. She first noticed it when she went outside to smoke a cigarette. It appeared to be coming from the buildings to the east of the food warehouse. The dispatcher further explained that it became more difficult to understand the caller because she was coughing so violently. Because your crew had previously preplanned many of the warehouses in that area, you decided to reference the copies of the preplans while en route to the location. There are two warehouses to the east, one housing lumber and building supplies, the other, a pool supply and recreation warehouse.

The dispatcher updates you on weather conditions while you are en route. It is 84° F, Humidity is 66%, and wind is out of the southeast at 12 mph. You arrive on the scene placing your unit upwind and uphill of the situation. A scan of the scene indicates what appears to be a white to light yellow vapor blowing across the parking lot to the south side. A closer look, using binoculars, indicates the vapor is seeping through the doors of the pool supply warehouse.

Scenario Questions

1. What should be your next function once you have pinpointed which warehouse the vapors are issuing from?

2. Knowing that a patient exists in the warehouse to the west, when should rescue be considered and how should it be accomplished?

3. To help identify the chemicals involved, what clues can you look for without placing yourself in danger?

4. To establish an initial hazard zone, what factors must be used to determine the size of the incident?

5. Several box truck trailers are in the fenced parking area and, with binoculars, you can see a yellow placard and a green placard. A red dangerous placard is also found on one truck. What might these placards indicate?

____

____

____

____

____

6. In addition to the DOT placards found on the trailers, the building has a multi-colored placard with numbers in each color. The blue quadrant indicates a 4, the red a 2, and the yellow a 3. The white section does not have any marks. What is this placard, and what degree of hazards are present?

____

____

____

____

____

7. You were just advised by your dispatch that an ambulance crew was sent to the location of the security guard. They approached the scene from downwind and are now inside the warehouse to the west and have advised their dispatcher that they, too, are having difficulty breathing, are coughing, and tearing eyes. They are unable to drive their vehicle and are requesting that the fire department hazardous materials team respond. What do you do to help coordinate this situation?

____

____

____

____

____

____

8. Since you now know that at least three patients exist, who should be informed?

____

____

____

____

____

9. Are there any special considerations for the care of these patients?

____

____

____

____

____

10. You determine that immediate treatment and transportation is a must since the closest hospital is 30 minutes away. You have been advised that a medical transport helicopter is en route and will be on the scene shortly. Is the use of a helicopter a viable solution for the long transport time?

____

____

____

____

____

Scene Organization and Standard Operating Guidelines

Objectives

Given a hazardous materials incident, you should understand the organization of the hazardous materials scene and identify stages of the event; the competencies necessary to safely work within the command structure of an incident in which patients are exposed to a chemical; and your obligations to the victim, to the team, and to command.

As a hazardous materials medical technician, you should be able to:

- Describe the overall priorities of a hazardous materials incident and the special considerations of each.
- List the three functional groups involved in most hazardous materials incidents.
- Describe the concerns of transporting a victim of a hazardous material exposure and identify the options for dealing with them.
- List the different types of decontamination and define each.
- List the steps that should be taken during site termination.
- Identify the differences between A, B, C, and D levels of protection.
- Define the three types of suit failure.
- List and define the seven stages of a hazardous materials incident as described in the chapter.

CONTENT QUESTIONS

1. What is the reason for an organized command structure? Does the command structure address the fact that the span of control is limited to a certain number of functions or persons? What is the ideal number needed to maintain the span of control?

____

____

____

____

____

2. On any hazardous materials incident where victims are involved, there are two types of victims. What are the two types and which type of victim would you think is the hardest to deal with? Why?

____

____

____

____

____

3. Emergency entry is identified as a possibility during a hazardous materials incident. When might emergency entry be considered? What factors may eliminate the option of emergency entry?

____

____

____

____

____

4. Draw a command structure indicating incident command, the command staff, and the functional groups. Indicate by name the persons in your organization who might be assigned these functions. Justify why these names were chosen.

____

____

____

____

____

5. Hazardous materials response falls under the operations section. Where would rehabilitation be positioned in the hazmat command structure?

6. A.P.I.E. is an acronym used to describe what? What do the letters stand for?

7. Seven steps have been identified as part of a hazardous materials incident. Investigation is the first step and must be started prior to advancing to the next. Name each of the steps and determine if the step must be started or completed prior to moving on to the next step. Justify your answer.

8. Why was the terminology Incident Management System chosen instead of Incident Command System for the book?

9. Reference for an incident begins and ends when?
   A. Before the incident takes place; when entry is made.
   B. Upon arrival; after exit from the hot zone.
   C. During preplanning; at site termination.
   D. Upon arrival; once the incident commander decides what to do.

10. What level of suit consists of full encapsulation with self-contained breathing apparatus inside? It is used in atmospheres that pose a threat through absorption, inhalation, and ingestion.

A. Level A B. Level B C. Level C D. Level D

11. Responder safety is the responsibility of:

A. Safety officer
B. Incident commander
C. Rehabilitation sector
D. Reference
E. All should have the first priority of responder safety

Answer questions 12–18 using one of the following stages.

A. Investigation stage
B. Planning stage
C. Preparation stage
D. Hazard mitigation stage
E. Cleanup
F. Site termination
G. Critique

12. Wind speed, direction, temperature, and even preliminary chemical information should be gathered during which stage?

____

____

____

____

____

13. Questions such as: What is the hazard? What will happen if nothing is done? What must be done? Are we capable of performing the needed tasks? are answered during what stage?

____

14. During what stage of the operation will the medical section establish the toxicity of the chemical and make preparations to treat entry team members if exposure exists?

____

15. Decontamination is found in what stage?

____

16. Post entry physicals are conducted to ensure that entry team members are able to return to work. These physicals are performed during the ______________ stage.

17. Several steps, such as advising the responsible party about the hazards that still may exist: advising as to the work completed for mitigation; providing a list of cleanup contractors and information contacts as needed; advising about any known legal requirements etc. should be done during the ________________ stage.

18. Identifying areas of the operation that need improvement or could be handled in a safer, or more efficient manner is done in the ________________ stage.

19. Define decontamination.

____

____

____

____

____

Match definitions 20–22 with the following terms.

A. Technical decontamination
B. Secondary decontamination
C. Gross decontamination

20. Usually involves removal of all of the clothing and a quick rinsing or washing of the skin: ________________

21. Used to decontaminate tools and equipment before returning it to service. Should never be done on personnel. May involve a neutralization process or special chemical washes: ________________

22. Mechanical removal or diluting of a chemical found on the skin or clothing (including suits) of an individual. Solution of choice for this procedure is usually a mild detergent and water: ________________

23. The incident management system allows the manager to
    A. let other commanders on the scene make decisions for him.
    B. add needed aspects of the incident and delete unnecessary ones.
    C. protect himself from legalities associated with incident command.
    D. cope with single engine responses.

24. Since the NFPA documents are consensus standards they
    A. only need to be followed if the department wishes to.
    B. can be totally ignored.
    C. are recognized as standards of practice and therefore may be used in court.
    D. must be followed to the letter as they are more enforceable than law.

25. The Incident Management System can be used for any emergency that
- **A.** exceeds the span of control.
- **B.** has only single unit responses.
- **C.** exceeds the span of control by 15 positions.
- **D.** involves only fire type situations.

26. Size-up
- **A.** is the gathering and organizing of all the known facts about the incident.
- **B.** starts even before the incident takes place.
- **C.** considers the time of day, weather, and location.
- **D.** is all of the above.

27. The primary goal of hazardous materials emergency response is to
- **A.** handle runoff and leaks.
- **B.** treat patients and avoid contamination.
- **C.** evacuate all endangered areas.
- **D.** limit the damage to life and property.

28. Although rapid response works for most fire scenes it is not normally recommended for hazardous materials scenes because
- **A.** hazmat scenes usually are over before the units arrive.
- **B.** the danger to responders may be so great that proper precautions are needed prior to entry.
- **C.** to mitigate the hazards involved in most fire situations, rapid actions are not necessary.
- **D.** hazardous materials spills and leaks will usually go away if left alone.

29. Functional groups in direct support of incident command are made up of
- **A.** finance, bureau, operations, PIO.
- **B.** PIO, billing, facilities, logistics.
- **C.** logistics, procurement, compensation, planning.
- **D.** planning, operations, logistics, finance.

30. Finance is further broken down into
- **A.** time, equipment, compensation, cost.
- **B.** equipment, medical, cost, compensation.
- **C.** logistics, time, cost, equipment.
- **D.** time, logistics, equipment, cost.

31. Logistics has two main branches called
- **A.** service, section.
- **B.** service, sector.
- **C.** service, support.
- **D.** support, search.

32. The operations section is responsible for
- **A.** command of the scene.
- **B.** coordination of the scene.
- **C.** the actual mitigation work.
- **D.** monies.

33. Hazardous Materials command is usually divided into three major groups. They are:

A. Medical, mitigation, legalities
B. Operations, safety, medical
C. Equipment, reference, safety
D. Reference, entry, safety

34. A PIO's job entails

A. keeping command posted on the efforts at the emergency.
B. identifying press so that command knows who to avoid.
C. distributing information to the new media about the incident.
D. informing the public in the area about the hazard so they can evacuate.

Use the definitions below to answer number 35 through 37.

35. Penetration is

____

36. Degradation is

____

37. Permeation is

A. The movement of a chemical through the protective material at the molecular level.
B. Physical movement of a chemical through the natural openings of a protective garment.
C. The physical destruction of a garment by temperature, chemical exposure, poor storage, etc.
D. A level of protection provided by a suit.

CONTENT APPLICATION #1

You have just found yourself in charge of a large working fire where hazardous materials may be a factor. The initial unit, dispatched to a trash fire, arrived on the scene to find a large dumpster fire and a tanker truck in close proximity to the scene also burning. The truck is parked next to an occupied strip mall. The engine officer who first arrived on the scene assumed command and asked for two full alarms. You happened to be only a block from the scene when you heard the dispatch. Upon your arrival, command was passed to you as identified in the department's SOPs. As incident commander, you must arrange a command structure. Without identifying the tactics you might use, draw out the command structure utilizing real personnel from your department or mutual aid departments. Place functions under each assigned unit. If a hazmat unit is needed, also identify the chain of command used within the hazmat sector and break down the structure to identify each function, group, and person. Be as specific as possible. Remember to limit the span of control of any unit or person. Be prepared to discuss your command structure in class.

CONTENT APPLICATION #2

Choose a recent hazardous materials incident in your jurisdiction (size is not a factor) and identify the command structure used there. If you are not familiar with the incident structure, draw one that would have worked for the incident. Does it matter how large or small the incident is for the command structure to work? Would the scene have worked better if the functions were divided into even smaller units? Explain.

CHAPTER 2 SCENARIO

You are the officer of an advanced life support (ALS) pumper and have just arrived on the scene of a large structure fire in a vacant warehouse on the riverfront. The warehouse, which has been vacant for at least 20 years, once housed fertilizers and pesticides for the local farming community. Because of a large amusement park built in the area and the booming growth of the tourist industry, the farms have virtually disappeared. In fact, just adjacent to the burning structure is an entertainment facility with a gambling riverboat tied up at the dock. The facility is filled with tourists ready to embark on their trip up the river. Many of them are outside watching the efforts of the fire department. When you refer to the preplan you find that a smaller detached structure contains abandoned tanks of anhydrous ammonia and pallets of ammonium nitrite. You also find that the property was targeted as a hazardous materials site and slated for cleanup during the next year. You request the hazmat team for their expertise in dealing with the additional hazards.

In addition to dealing with the large population of tourists, you now notice the Eye Of the Sky helicopter over the fire, fanning the smoke in different directions and the On the Spot News van setting up for a live broadcast just across the street from the fire.

Scenario Questions

1. Who makes up the initial command structure?

2. What functional groups would need to be immediately set up for this incident?

3. Would you order the news agencies out of the area or off of the scene? Where and why?

4. When the hazmat team arrives, where in the command structure would they belong?

5. What means can the hazmat team use to identify the hazards present?

6. Once identified, what type of research is necessary to properly control the incident?

7. If the firefighters became grossly contaminated, what additional functions will be necessary prior to treatment? Transportation?

____

____

____

____

____

8. What would be the main concerns of the hazmat team upon their arrival?

____

____

____

____

____

9. Would you think that runoff from this fire is a concern?

____

____

____

____

____

10. Is a decontamination area necessary? Is this a function to be provided by the hazmat team? If so why or why not?

____

____

____

____

____

Chemical Behavior

Objectives

Given a list of chemical reference criteria, you should be able to define, identify, and appropriately describe their importance as they relate to risk assessment, health, and safety surrounding a hazardous materials incident. You should be able to describe the physical, chemical properties, concentrations, strengths, structure, and form of the hazardous material primarily as they relate to product identification, personnel protection criteria, decontamination procedures, and treatment protocols.

As a hazardous materials medical technician, you should be able to:

- Describe the relative chemical and physical properties that may present as a health risk to the emergency worker.

  Melting point
  Boiling point
  Density
  Solubility
  Vapor pressure
  Appearance
  Expansion ratios
  Temperature
  Air reactivity
  Chemical reactions
  Inhibitor
  Salt
  Saturated hydrocarbon
  Aromatic hydrocarbon
  Specific gravity
  Vapor density
  Viscosity
  Ignition temperature
  Flammable explosive limits
  Flash point
  Reactivity
  Combustion
  Catalyst
  Critical temperature and pressure
  Polymerization
  Nonsalt
  Unsaturated hydrocarbon

- Define the atomic structure and how it relates to the risk assessment.
- Describe how bonding can give clues on chemical behavior.
- Identify a chemical and its risk assessment based upon the symbols, and/or formula.

- Describe the risk assessment of a chemical based upon its location on the periodic chart.
- Identify the differences in bonding.
- Describe the principles of bonding and its relevance to the risk vs. gain process.
- Describe the principles of the gas laws and their affect in decision making.
- Describe how the rates of reaction can affect the risk assessment.
- Identify the principles of acid–base reactions in terms of:

  Concentration | Strength | pH

- Identify the hazards and risk assessment of the following organic families:

Alcohols	Amines	Esters
Aldehydes	Carboxylic acid	Ethers
Alkyl halides	Ketones	

- When given a scenario, the risk assessment, health effects and possible chemical reactions shall be identified utilizing:

Technical resources	Technical information centers	NAERG handbook
Material safety data sheets	Interpretive data	Reference manuals

CONTENT QUESTIONS

1. The study of chemistry is divided into how many separate but closely related branches?
   - **A.** One, chemistry
   - **B.** Two, inorganic chemistry and organic chemistry
   - **C.** Three, inorganic, organic and biochemistry
   - **D.** Four, inorganic, organic, toxicological chemistry

2. There are on average, according to this textbook, how many naturally occurring elements?

   A. 89 **B.** 94 **C.** 102 **D.** 92

3. All matter consists of three basic states. What are these three states?
   - **A.** Volume, mass, and liquid
   - **B.** Solid, liquid, and gas
   - **C.** Atoms, elements, and molecules
   - **D.** Liquid, elements, and gas

4. The tiny discrete particles are called
   - **A.** atoms.
   - **B.** molecules.
   - **C.** elements.
   - **D.** compounds.

5. Within the basic fundamental elements stated in the answer in question 4, these tiny discrete particles contain smaller units of structure. What are these units called?
   - **A.** Neutrinos, particles, and electrons
   - **B.** Alpha, beta, and gamma
   - **C.** Neutrons, protons, and electrons
   - **D.** Neutrons, protons, and photons

6. The basic unit identified on the periodic chart is called the
   - **A.** proton.
   - **B.** neutron.
   - **C.** element.
   - **D.** compound.

7. Compounds, molecules, and elements have properties that are characteristic to that particular material. These properties are known as
   - **A.** chemical properties.
   - **B.** physical properties.
   - **C.** reaction properties.
   - **D.** both A and B.

8. What best defines a compound?
   - **A.** Further decomposition is not possible
   - **B.** Formed by the union of the respective elements
   - **C.** Union of the elements creating a chemical unit
   - **D.** Compounds are different from molecules

9. What type of property is it when we have qualitative numbers?
   - **A.** Chemical
   - **B.** Particle
   - **C.** Physical
   - **D.** Empirical

10. The above question stated a set of qualitative numbers. These numbers are gathered under what is called
    - A. Standard atmospheric conditions of pressure
    - B. Standard pressure and temperature
    - C. Standard conditions of temperature
    - D. Standard temperature and pressure

11. Define the standard conditions.

____
____
____
____
____

12. Define the conditions that are abbreviated o.c. and c.c.

____
____
____
____
____

13. Give examples of physical properties.

____
____
____
____
____

14. The form of the hazardous material is a description of what property?
    - A. Vapor pressure
    - B. Melting point
    - C. Boiling point
    - D. Appearance

15. The ability of the material to change from a solid to a liquid is a definition of
    - A. vapor pressure.
    - B. melting point.
    - C. boiling point.
    - D. appearance.

16. The temperature at which the atmospheric pressure and the vapor pressure of a liquid are equal is a definition of
    - A. vapor pressure.
    - B. melting point.
    - C. boiling point.
    - D. appearance.

17. Density is defined best by which statement?
    A. Hardness of a substance
    B. Quantity of a substance
    C. Relationship between mass and volume
    D. Mass per unit volume which for solids and liquids is expressed as grams per centimeter cubed

18. A comparison of weight between water and the material being tested is called
    A. specific density.
    C. specific gravity.
    B. density.
    D. vapor density.

19. Define how the above principle is used in the hazardous materials incident.

____

____

____

____

____

20. A comparison of a gas or vapor to the ambient air is called
    A. specific density.
    C. specific gravity.
    B. density.
    D. vapor density.

21. Define how the above principle is used in the hazardous materials incident.

____

____

____

____

____

22. ____________________ is the ability of a material to blend uniformly within another material.

    Certain materials are ____________________ in any proportion, while others are not.

    Being ____________________ refers to the ability to dissolve into a uniform mixture.

23. The above blend is called a ____________________ .

    The material that is in the greatest amount is called the ____________________ while the material that is in lesser amount (usually the additive) is called the ____________________.

24. Define how the above principle is used in the hazardous materials incident.

____

____

____

____

____

25. ____________________ has a lot to do with solubility, ____________________ substances have a positive end and a negative end, while ____________________ substances do not have an electronegative or electropositive end.

26. Define how the above principle is used in the hazardous materials incident.

____

____

____

____

____

27. A molecule has escaped the liquid form and has been changed into a molecule traveling through the air space giving it a gaseous state. This is called

A. vapor density.
B. vapor pressure.
C. vapor temperature.
D. vapor change.

28. The minimum temperature under which a liquid will give off vapors to form an ignitable mixture in air is called

A. vapor pressure.
B. fire point.
C. ignition temperature.
D. vapor point.

29. When a gas is released into the environment, the wind, ambient temperature, and topography play roles in the chemical's ability to ignite. It must find an available ignition source. If we could measure the gas, as it is traveling through the air, we would see three distinct areas of concentration. These areas are called

A. flammable ranges.
B. explosive limits.
C. flammable limits.
D. all the above.

30. ____________________ is a measure of flow. It determines the thickness of a liquid, or how well it flows.

A low ____________________ liquid will flow like water.

The lower the ____________________ the higher the tendency for the liquid to spread.

On the other hand the higher the ____________________ the slower the spread.

31. Define how the above principle is used in the hazardous materials incident.

____

____

____

____

32. If the product is going from the liquid phase into a solid phase we call it:

____

33. If the chemical is moving from a solid to the liquid phase we call it:

____

34. ______________________ is the minimum temperature that a material must be raised to in order for it to ignite and sustain combustion.

A. Heat of combustion
B. Heat of vaporization
C. Flash point
D. Ignition temperature

35. The particle that revolves about the center portion of an atom is called a (an)

A. electrino.
B. positron.
C. negatron.
D. electron.

36. The center of an atom has two particles that make up the nucleus. They are called

A. protons.
B. electrons.
C. neutrons.
D. A and C.

37. All atomic nuclei contain an equal number of protons in relation to the number of electrons that are in orbit.

A. True
B. False

38. The atomic number is used to "catalog" the elements.

A. True
B. False

39. Atoms of the same element may have a different number of neutrons and protons within the nucleus. These classes of the same atom are called

A. isomers.
B. isotopes.
C. ionic.
D. ions.

40. How many electrons are in the outer shell in group one?

A. 1 **B.** 2 **C.** 3 **D.** 4

41. How many electrons can it accept or lose?

A. 4 **B.** 5 **C.** 6 **D.** 7

42. How many electrons are in the outer shell in group two?

A. 1 **B.** 2 **C.** 3 **D.** 4

43. How many electrons can it accept or lose?

A. 4 **B.** 5 **C.** 6 **D.** 7

44. How many electrons are in the outer shell in group 13?
A. 1 **B.** 2 **C.** 3 **D.** 4

45. How many electrons can it accept or lose?
A. 4 **B.** 5 **C.** 6 **D.** 7

46. How many electrons are in the outer shell in group 14?
A. 1 **B.** 2 **C.** 3 **D.** 4

47. How many electrons can it accept or lose?
A. 4 **B.** 5 **C.** 6 **D.** 7

48. How many electrons are in the outer shell in group 15?
A. 4 **B.** 5 **C.** 6 **D.** 7

49. How many electrons can it accept or lose?
A. 4 **B.** 5 **C.** 6 **D.** 7

50. How many electrons are in the outer shell in group 16?
A. 4 **B.** 5 **C.** 6 **D.** 7

51. How many electrons can it accept or lose?
A. 4 **B.** 5 **C.** 6 **D.** 7

52. How many electrons are in the outer shell in group 17?
A. 4 **B.** 5 **C.** 6 **D.** 7

53. How many electrons can it accept or lose?
A. 4 **B.** 5 **C.** 6 **D.** 7

54. Give the chemical names for the following abbreviations.
A. Li ______________________________
B. Na ______________________________
C. K ______________________________
D. Rb ______________________________
E. Cs ______________________________
F. Fr ______________________________

55. What is the name for the first group?
A. Alkali metals
B. Alkaline earth metals
C. Transitional metals
D. Nitrogen family

56. This group is ______________________ in the pure form.

They ______________________ an electron easily due to the fact of the one electron in the outer orbital.

Because of this, rarely do you find any of this group in the raw form in nature.

The reactions of these elements are so violent that the commercial uses are limited to ______________________ and ______________________.

57. ______________________ and ______________________ although primarily used chemicals in this group, have their own health hazards.

They produce ______________________.

These ______________________ can produce severe burns and irritate the ______________________.

58. Describe the general hazards of the alkali metals.

____

____

____

____

____

59. What is the name for the second group?

A. Alkali metals
B. Alkaline earth metals
C. Transitional metals
D. Nitrogen family

60. Give the chemical names for the following abbreviations.

A. Be ____
B. Mg ____
C. Ca ____
D. Sr ____
E. Ba ____
F. Ra ____

61. ______________________ is radioactive. As the element degenerates, it is transformed into a lower atomic weighted element. This element is a strong radiological hazard. All precautions should prevail around this element.

______________________ has the potential to emit ______________________, ______________________, particles and ______________________ radiation.

62. This group, as with group one, is also reactive, due to the valence of ___________.

Here we have ______________________ electrons in the outer shell. Because of this they react with ______________________, releasing ______________________, a fuel source.

63. Describe the chemical hazards of CaO, calcium oxide or quicklime.

____

____

____

____

____

64. Discuss the general hazards for this chemical group.

____

____

____

____

____

65. What is the name given for the next ten groups?

A. Alkali metals
B. Alkaline earth metals
C. Transitional metals
D. Nitrogen family

66. Why do we concern ourselves with these elements medically?

____

____

____

____

____

67. Give the chemical names for the following abbreviations.

A. Hg ______________________________
B. Cd ______________________________
C. Cr ______________________________
D. Ti ______________________________
E. Zr ______________________________
F. Zn ______________________________

68. Describe the general hazards of the transitional metals.

____

____

____

____

____

69. Give the chemical names for the following abbreviations.

A. B ______________________________
B. Al ______________________________
C. Ga ______________________________
D. In ______________________________
E. Tl ______________________________

70. Describe the general hazards of group 13.

____

____

____

____

____

71. Give the chemical names for the following abbreviations.

A. C _______________

B. Si _______________

C. Ge _______________

D. Sn _______________

E. Pb _______________

72. Describe the general hazards of group 14.

____

____

____

____

____

73. Give the chemical names for the following abbreviations.

A. N _______________

B. P _______________

C. As _______________

D. Sb _______________

E. Bi _______________

74. _______________ is the only element within this group that is a gas. The rest are in solid form.

75. _______________ because of its naturally occurring state, can be used to form oxides.

These oxides are called _______________ .

Once in this state it becomes a gas, creating a very toxic compound.

The _______________ are severe _______________ that if inhaled can cause severe _______________.

The reaction disrupts the _______________ in the lungs.

76. Name four areas that the decontamination team should pay particular attention to.

1. ____________________

2. ____________________

3. ____________________

4. ____________________

77. ____________________ and the ____________________ can form to produce compounds called ____________________.

These, too, should be avoided. If in solid form, such as powders or granules, the material will irritate ____________________.

If in a vapor state, ____________________ should be the number one priority.

In these cases ____________________ is observed due to the tissue damage in the lungs.

78. Agricultural poisons and insecticides are all examples of ____________________ uses.

79. Describe the general hazards of group 15.

____

____

____

____

____

80. Give the chemical names for the following abbreviations.

A. O ____________________

B. S ____________________

C. Se ____________________

D. Te ____________________

E. Po ____________________

81. Name the following compounds.

A. A toxic gas from the decay of organic compounds

B. A refrigerant in old systems

C. Widespread in chemical processes

82. Name a chemical that can develop “garlic breath.”

____

83. Describe the general hazards of group 16.

____

____

____

____

____

84. What is the name given for group 17?

A. Alkali metals
B. Alkaline earth metals
C. Transitional metals
D. Halogens

85. Give the chemical names for the following abbreviations.

A. F ______________________________
B. Cl ______________________________
C. Br ______________________________
D. I ______________________________
E. At ______________________________

86. Which of the following is a true statement?

A. Alkali earth metals do not react with water.
B. Nitrogen oxides are non-toxic.
C. All halogens are harmful.
D. All the above.

87. How many electrons can an atom accept or lose?

A. 7 **B.** 5 **C.** 3 **D.** 1

88. Which of the following choices is the most electronegative element?

A. O **B.** Br **C.** F **D.** N

89. Chlorine and its by-products are extensively used as disinfectants for swimming pools, hot tubs and Jacuzzis. The two most common are ______________ and ______________. These two chemicals are ______________ with each other and with other pool treatments. If they come in contact with ______________, ______________, ______________, ______________ or ______________, the heat produced can start a reaction that can result in spontaneous combustion.

90. Describe the oxidative hazards of this chemical group.

____

____

____

____

____

91. Describe the general hazards of this group.

____

____

____

____

____

92. What is the name given for group 18?

A. Alkali metals
B. Alkaline earth metals
C. Noble gases
D. Halogens

93. Give the chemical names for the following abbreviations.

A. He ______________________
B. Ne ______________________
C. Ar ______________________
D. Kr ______________________
E. Xe ______________________
F. Rn ______________________

94. The reason for stability is due to the electron configuration in the outside shell. All of these elements contain ______________ electrons, thus making them stable.

95. They are ALL able to displace ____________ and can cause ____________ .

96. Elements that have had their outermost electron stripped from their orbits have one or two fewer electrons and are called ______________ , and possess a greater positive charge. While on the other hand, we can see cases that have one or two more electrons added to the outside shell. These are called ____________. Collectively this group of atoms is called ____________.

97. Name the following –1 anions.

A. H ______________________
B. F ______________________
C. Cl ______________________
D. Br ______________________
E. I ______________________

98. Name the following –2 anions.

A. O ______________________
B. S ______________________
C. Se ______________________

99. Name the following oxygenated inorganic –1 ions.

A. NO_2 ______________________
B. ClO_2 ______________________

100. Name the following oxygenated –1 ions.

A. NO_3 ______________________

B. ClO_3 ______________________

101. Name the following oxygenated –2 ions.

A. SO_4 ______________________

B. CrO_4 ______________________

C. CO_3 ______________________

102. Name the following –3 oxygenated ions.

A. PO_4 ______________________

B. BO_3 ______________________

103. Name the following –1 radicals.

A. ClO_4 ______________________

B. ClO_3 ______________________

C. ClO_2 ______________________

D. ClO ______________________

104. Name the following –2 radicals.

A. CO_4 ______________________

B. CO_3 ______________________

C. CO_2 ______________________

D. CO ______________________

105. Name the following –3 radicals.

A. PO_5 ______________________

B. PO_4 ______________________

C. PO_3 ______________________

D. PO_2 ______________________

106. Identify the type of ion (positive or negative) and the abbreviation of the following.

A. Mercurous ______________________

B. Cuprous ______________________

C. Manganic ______________________

D. Ferrous ______________________

E. Stannic ______________________

F. Cupric ______________________

G. Mercuric ______________________

H. Ferric ______________________

I. Stannous ______________________

J. Manganous ______________________

107. An ionic bond between two atoms can be best described as

A. an atom of low electronegativity and high ionization.

B. an atom of high electronegativity and high ionization.

C. an atom of high electronegativity and low ionization.

D. an atom of high electronegativity and high ionization.

108. ______________________ is an element's ability to attract a pair of electrons.

109. How is this group formed and what are the hazards?

____

____

____

____

____

110. How does the naming configuration identify the ion or radical?

____

____

____

____

____

111. Name the group that has a metal and oxygen in attachment, and the ending assigned. ____

112. Name the group that has the O_2 radical and its hazards.

____

____

____

____

____

113. Name the group that is identified by the –OH radical its hazards.

____

____

____

____

____

114. Describe the oxygenated salts and their hazards.

____

____

____

____

115. Bonds are called ____________________ when the electrons are shared.

116. What is the maximum number of electrons that revolve around the nucleus in the outermost shell?

A. 7 **B.** 6 **C.** 3 **D.** 8

117. What are the exceptions to the above question?

____

____

____

____

118. Describe ionization potential and where this theory is applied.

____

____

____

____

119. Discuss how we can rapidly identify an element's ionization potential and atomic radius.

____

____

____

____

____

120. Discuss how ionic bonds and covalent bonds are different.

____

____

____

____

____

121. What theory best describes the movement of gases?
  A. Henry's theory of solutions
  B. Dalton's theory of partial pressures
  C. Kinetic molecular theory
  D. Boyle's theory

122. Discuss critical temperature and critical pressure.

____

____

____

____

123. In terms of pressure, what numbers can be considered as equals?

____

____

____

____

124. Describe each of the gas laws.

Boyle's Law

____

____

____

____

Charles's Law

____

____

____

____

Dalton's Law

____

____

____

____

Henry's Law

____

____

____

____

____

125. Describe the movement of heat.

____

____

____

____

____

126. Apply this to the medical aspects of a hazardous materials incident.

____

____

____

____

____

127. Describe the calorie and the two types of reactions it measures.

____

____

____

____

____

128. Describe ignition temperature.

____

____

____

____

____

129. Describe activation energy.

____

____

____

____

____

130. Describe an effective collision and how this may occur.

____

____

____

____

____

131. The acid–base reaction can be described as a chemical's ability to produce what two molecules?

A. Water, H_2O

B. Hydronium ion, H_3O^+

C. Hydroxide ion, OH^-

D. Hydrogen ion, H

132. What are the ranges of pH?

A. 1–7 acidity, 7.1–14 alkalinity

B. 0–6.9 acidity, 7.1–14 alkalinity

C. 0–7 acidity, 7–14 alkalinity

D. 0–6.8 acidity, 7.2–14 alkalinity

133. What reference scale may textbooks utilize and how are these numbers used?

____

____

____

____

____

134. Name the group that is assigned to the number carbons stated below.

A. One carbon ______________________________

B. Two carbons ______________________________

C. Three carbons ______________________________

D. Four carbons ______________________________

E. Five carbons ______________________________

F. Six carbons ______________________________

G. Seven carbons ______________________________

H. Eight carbons ______

I. Nine carbons ______

J. Ten carbons ______

K. Eleven carbons ______

L. Twelve carbons ______

135. Name the organic family and general formula for a single, double, and triple bond.

______

______

______

______

______

136. Utilize the IUPAC common naming system for the following compounds.

A. CH_3-CH_2-CH_2-CH_2-CH_2-CH=CH_2 ______

B. CH_2-CH=CH-CH=CH-CH_3 ______

C. CH_3-CO-CH_2CH_3 ______

D. $CH_3CH_2CH_2CH_2$-CO-CH_2CH_3 ______

E. $CH_3CH_2CH_2$-CO-$CH_2CH_2CH_3$ ______

F. CH_3-C=C-CH_2-CH_3 ______

G. $CH_3CH_2NH_2$ ______

H. CH_3-OH ______

I. CH_3-C-O-O-CH_3 ______

J. CH_2OH-CH_2OH ______

K. CH_3-CH=CH-CH-CH_2-CH-CH_2-CH_3

| |

CH_3 CH-CH_3 ______

137. Draw the following compounds.

A. p-Chlorophenol

B. 3-Chloroheptane

C. 2-methyl-4-ethyl-2-octene

D. Pent-2-yne

E. trans- Dichloroethlene

F. Styrene

G. Acrylaldehyde

H. Valeric Acid

I. 1,2-Ethanediol

J. Ethyl acetate

K. Heptane

L. 2,4-Hexadiene

M. Acetaldehyde

N. MEK

O. Vinyl methyl ether

P. 2-Propenamine

Q. Dimethylamine

R. Aluminum trichloride

S. Sulfur trioxide

T. Chlorous acid

U. Benzonitrile

V. Nitrocellulose

W. Trinitrophenol

X. Dinitrogen tetroxide

Y. Aniline

138. Describe the orientation of compounds that are mirror images.

139. Describe the hazards of each of the following classifications:

A. Alcohols

B. Phenols

C. Ketones

D. Esters

E. Aldehydes

F. Alkyl halides

G. Amines

______

______

______

H. Ethers

______

______

______

140. Name the following compounds and their representative group:

A. $HClO_4$ ______
B. CH_3-O-O-CH_2CH_3 ______
C. $SbCl_5$ ______
D. SO_3 ______
E. $C_6H_5CH_3(NO_2)_3$ ______
F. Thiols R-SH ______
G. MnO_2 ______
H. CH_3CN ______

CONTENT APPLICATION

You are preplanning several occupancies and confined spaces within your jurisdiction. During your investigation you have created a listing of the chemicals that are common throughout the response area. For each chemical, identify the family of chemicals that it comes from, possible synonyms using nomenclature, hazards that each family will present with, and the possible medical concerns.

A. Compound 1080 ______

______

B. Sodium chlorite ______

______

C. Mercaptan ______

______

D. Cement ______

______

E. Calcium peroxide ______

______

F. Potassium sulfide______

______

G. Triethylamine ______________________________

____

H. Hypochlorous acid ______________________________

____

I. Lye ______________________________

____

J. Hex-2-yne ______________________________

____

K. Chlorous acid ______________________________

____

L. Calcium oxide______________________________

____

M. Ammonia ______________________________

____

N. Methyl acetate ______________________________

____

O. Sulfur trioxide ______________________________

____

P. Cellulose nitrate ______________________________

____

Q. Cupric chloride ______________________________

____

R. Acetaldhyde ______________________________

____

S. Acrylonitrile ______________________________

____

T. Styrene ______________________________

____

U. Sodium perchlorate ______________________________

____

V. Vinyl trichloride ______________________________

____

W. Aluminum trichloride ______________________________

____

X. Barium cyanoplatinite ______________________________

____

CHAPTER 3 SCENARIO

During a company survey of an occupancy, your engine company finds the storage of some chemicals. As a part of department protocol, the hazmat team is called to the facility to research some of the found chemicals. Because of another incident occurring in the south part of town, the hazmat team is riding one firefighter short. The hazmat captain knows that you just completed a course on Hazardous Materials Medical Considerations and asks you to perform the research. You have available to you the MSDS, commonly found reference texts, and computer access through a local emergency planning agency.

Scenario Questions

1. What chemical and physical properties should you initially research?

2. Knowing the chemical and physical properties, how would the temperature affect a variety of inorganic, organic, ionic or covalently bonded chemicals?

3. What would water do to your inorganic chemicals, and what category should you be concerned with?

4. If a cryogenic was involved, what are the considerations of the above discussions?

5. Discuss the inorganic compounds in terms of fire potential and health hazards.

6. Discuss the organic compounds in terms of fire potential and health hazards.

7. What other categories of compounds should you identify?

Essentials of Toxicology

Objectives

When presented with a possible toxic exposure, the student shall be able to work through the risk assessment as it relates to the acute toxic event. In addition, the numbers that are referenced as levels of possible concern should be understood for their scientific values rather than their absolute virtues.

As a hazardous materials medical technician, you should be able to:

- Explain the differences between toxicity and toxic.
- Discuss the role metabolism has within the exposure event.
- Discuss the four basic factors of exposure.
- Explain the principles of toxicology and how they relate to the exposure individual:

  Graded response
  Quantal response
  Dose response
  Effective concentration
  Routes of exposure
  Combined effects
  Phase I and II reactions
  Variables affecting toxic levels
  Acute exposure
  Subchronic exposure
  Chronic exposure

- Define the following toxicological terms:

  TLV
  TLV-TWA
  TLV-STEL
  TLV-EL
  TLV-s
  TLV-c
  IDLH
  PEL
  CL
  PK
  LClo
  LDlo
  LC_{50}
  LD_{50}
  LCT_{50}
  MAC
  RD_{50}
  WEEL

- Discuss the one basic truth about threshold limit values.
- Discuss the margin of safety.

CONTENT QUESTIONS

1. What educational discipline does toxicology fall under?
   A. Biology **B.** Toxicology **C.** Pharmacology **D.** Chemistry

2. The reaction that a chemical undergoes within an organism is referred to as biometabolism.
   A. True
   B. False

3. Match the following:
   A. Toxicity
   B. Toxin
   C. Contaminated
   D. Toxic exposure

   1. contact with the toxin
   2. the concentration of a dose that causes a response
   3. the ability of a chemical or substance to cause an injury or harm
   4. agent that can cause potential harm

4. What three questions should responders concern themselves with?
   A. ______________________________
   ____
   B. ______________________________
   ____
   C. ______________________________
   ____

5. An organism's ability to utilize foods for fuel is called
   A. biochemistry.
   B. phase I reactions.
   C. metabolism.
   D. Both C and B.

6. Match the following.
   A. Carbohydrates
   B. Proteins
   C. Fat

   1. compounds utilized by the cells to establish enzymes and cell structure
   2. chemical compounds called glucose
   3. carbohydrate stores

7. ____________________ is sometimes broken down, and this process is called ____________________.
   A. Proteins / anabolism
   B. Acids / anabolism
   C. Glucose / catabolism
   D. Proteins / catabolism

8. ________________ are built up and this process is called ________________.
   A. Proteins / anabolism
   B. Acids / anabolism
   C. Glucose / catabolism
   D. Proteins / catabolism

9. The liver, kidneys, and lungs are involved with the detoxification of a chemical in the general sense. What two basic scenarios can occur?

   1. ______________________________

   2. ______________________________

10. Describe liver metabolism and the complexities as presented in the textbook.

11. Describe Phase I reactions.

12. Describe Phase II reactions.

13. Discuss how polarity affects the toxic event.

14. How do the lungs affect the detoxification process?

15. Describe how chemical qualities influence the toxic character of a particular compound.

16. List the four basic truths that determine the toxic event response.

1.

2.

3.

4.

17. What is the response called that is represented by the number of receptor sites that are bound?
    A. Quantal response
    B. Dose response
    C. Graded response
    D. Effective concentration

18. The point at which a chemical has produced an effect is called
    A. quantal response.
    B. dose response.
    C. graded response.
    D. effective concentration.

19. The all or none response is called
    A. quantal response.
    B. dose response.
    C. graded response.
    D. effective concentration.

20. There are five groups that the body is viewed as having. Match the components with the group.
    A. Lung group ____________
    B. Vessel group ____________
    C. Blood group ____________
    D. Muscle group ____________
    E. Fat group ____________

    1. the brain
    2. muscle and skin
    3. respiratory tree
    4. composed of the adipose tissue and bone marrow
    5. liver, kidneys, heart, and the gastrointestinal tract

21. All of the groups have one thing in common. This mutual property is
    A. oxygenation.
    B. blood perfusion.
    C. partial pressure.
    D. metabolism.

22. As the individual assimilates the chemical, it will be equally distributed throughout the bloodstream. Through this equal distribution, the chemical's physical and chemical properties will target a particular organ. Discuss this in terms of polarity and solubility.

    ____

    ____

    ____

    ____

23. The overall observable reaction is called
    A. effective response.
    B. lethal concentration.
    C. dose response.
    D. effective concentration.

24. If one chemical toxifies a small percentage of the population, and one chemical insults a large percentage, discuss the differences of each chemical.

    ____

    ____

    ____

    ____

25. Discuss how standard deviation affects the interpretation of toxicological numbers.

____

____

____

____

26. The toxic levels of a chemical are commonly found

A. at the LEL.
B. within the flammable limits.
C. at the UEL.
D. below and slightly above the LEL.

27. Name the four routes of exposure.

A. ____
B. ____
C. ____
D. ____

28. The following is a numerical representation of combined effects. Match each effect with its numerical representation.

A. 1+1 = 2 ____
B. 1+1 = 3 ____
C. 1+0 = .5 ____
D. 1+0 = 2 ____

1. Synergism
2. Antagonism
3. Additive
4. Potentiation

29. Utilizing the above names for combined effects, place them with the appropriate definition.

A. The combined effects are more severe than the individual chemicals. In this case, we may have a chemical that by itself is moderately toxic, however, in combination with another, the toxic qualities can be enhanced. ____________________

B. A chemically inactive species acts upon another chemical, which enhances the chemically active substance. ________________

C. Some chemicals that are different in chemical structure (shape and polarity) may have the same physiological response in the organism. ________________

D. The combined effects cancel the effects of each other, decreasing the toxic event. What actually happens is that one of the chemicals is acting to decrease the effects of the other chemical. ____________________

30. Name the variables that can affect toxic levels.

____

____

____

____

31. Describe briefly each variable and how it can affect the dose response.

A. ______________________________

B. ______________________________

C. ______________________________

D. ______________________________

E. ______________________________

F. ______________________________

G. ______________________________

____

____

____

____

32. The following are abbreviations of specific toxicological values. Name them and the associated agency.

			Answer Column A	Answer Column B
A. PEL	________	________	1. Exposure limits	a. AIHA
B. MAK	________	________	2. Maximum allowable concentration	b. ACGIH
C. TLV	________	________	3. Threshold limit values	c. OSHA
D. EL	________	________	4. Permissible exposure limits	d. GRS
				e. NIOSH

33. Name the five conditions for which the TLV was intended.

A. ______________________________

____

____

____

B. ______________________________

____

____

____

C. ______________________________

____

____

____

D. ______________________________

____

____

____

E. ______________________________

____

____

____

34. Match the following definitions to the appropriate toxicological term.

1. NIOSH TLV-TWA ____________
2. STEL ____________
3. TLV-TWA ____________
4. EEL ____________
5. TLV-c ____________
6. TLV-s ____________

A. Is based upon an 8-hour day, 5 days a week to give a 40-hour work week.
B. Identifies that the material is absorbed through the skin.
C. Is based upon a 10-hour work day 4 days a week, to give the 40-hour work week.
D. Occurs for only 15 minutes and is not repeated more than 4 times a day. Each 15-minute exposure event is interrupted by a 60-minute non-exposure environment.
E. IDLH (Immediately Dangerous to Life and Health).
F. c denotes ceiling levels.

35. Match the following ACGIH values.

A. TLV-TWA ____________
B. TLV-STEL ____________
C. TLV-s ____________
D. TLV-c ____________
E. TLV-EL ____________

1. An average exposure not to exceed the published 8-hour TWA. This will not occur for more than 30 minutes on any work day.
2. Identifies a material that is absorbed through the skin.
3. Fifteen minute excursion in which the worker is exposed to the chemical continuously.
4. A ceiling level.
5. An 8-hour a day 40-hour work week with repeated exposure without any adverse effects.

36. Match the following OHSA and NIOSH values.

A. IDLH ____________
B. PK ____________
C. PEL ____________
D. REL ____________
E. CL ____________

1. This has the same meaning as TLV-TWA.
2. This is similar to the TLV-c.
3. This is the same as PEL, and TLV-TWA.
4. The maximum airborne contamination that an individual could escape from within thirty minutes without any side effects.
5. Is different depending on the testing agency, however, it is the apogee of the daily allowable limit.

37. Define IDLH in detail.

____

____

____

____

____

38. 50% of the total population under study died. How is this interpreted in the terminology? ______________________________

____

____

____

____

39. Match the following.

A. LClo ____________

B. LDlo ____________

C. LC_{50} ____________

D. LD_{50} ____________

E. LCT_{50} ____________

F. MAC ____________

G. RD_{50} ____________

1. The lowest concentration of airborne contaminates that caused injury.

2. 50% of the test population died from the introduction of this airborne contaminate.

3. A statistically derived LC_{50} (LDT_{50} statistically derived lethal dose).

4. A 50% calculated concentration of respiratory depression (RD = respiratory depression of 50% of the observed population), towards an irritant, over a 10–15-minute time frame.

5. The lowest dose (solid/liquid) that caused an injury.

6. This is the maximum allowable concentration (European).

7. Fifty percent of the tested population died from the introduction of this chemical which may be a solid, liquid, or gas.

40. Describe the acute exposure.

____

____

____

____

41. Describe the subchronic exposure.

42. Describe the chronic exposure.

43. Describe the margin of safety.

CONTENT APPLICATION

The following is a list that was generated while preplanning your response jurisdiction. Research the toxicological data for each chemical:

A. 1,2-ethanediol

B. Acrylaldehyde

C. Dimethylamine

D. Dinitrogen tetroxide ______________________________

____

____

____

E. Benzonitrile ______________________________

____

____

____

F. Calcium nitride ______________________________

____

____

____

G. Cesium hydroxide ______________________________

____

____

____

H. p-chlorophenol ______________________________

____

____

____

I. Ethylamine ______________________________

____

____

____

J. Methylphenyl ether______________________________

____

____

____

K. Perchloric acid ______________________________

____

____

____

L. Cyclonite ______________________________

____

____

____

M. Arsenic oxide ______________________________

____

____

____

N. Sodium chlorate______________________________

____

____

____

O. Butyl ethyl ketone ______________________________

____

____

____

P. Vinyl trichloride ______________________________

____

____

____

Q. Phosphorous trichloride ______________________________

____

____

____

R. Ethyl mercaptan ______________________________

____

____

____

S. Dimethyl peroxide ______________________________

____

____

____

T. Barium peroxide ______________________________

____

____

____

U. Nitrogen ______________________________

____

____

____

By utilizing the information presented in the appendix apply a risk vs. benefit analysis of each chemical presented above. Consider Nitrogen, Ethylamine, Ethyl mercaptan, and Methyl isocyanate within an occupancy with potential victims. What is your analysis?

CHAPTER 4 SCENARIO

You have been called by a neighboring hazardous materials team that is on the scene of several spilled chemicals. Because of reduced manpower, they are overwhelmed in reference sector. They have called you to provide the necessary medical research of the hazardous materials scene.

As you are recording the information, you recognize a few of the chemicals. One in particular has a very shallow toxicological curve, which occurs over large animal populations. You further assess that most of the chemicals are metabolized during phase one into highly polar and soluble products.

During the physical description of the incident, your cellular phone connection becomes full of static, and your connection is lost. Your only piece of solid information is that the incident is occurring in a populated area of town. You know the members of the other team and feel confident that they will call or utilize another method of communication shortly. In an attempt to ready yourself for the research task, you start by writing down a set of questions. With the addition of information as it becomes available you will base your research upon these questions.

SCENARIO QUESTIONS

1. What form of matter is the material in?

____

____

____

____

____

A. How will temperature affect the chemical and physical properties?

B. What are the potential routes of exposure?

C. What are the potential target organs that could be involved?

2. If several chemicals are involved, how could it affect the potential exposure?

3. You remember that one of the hazmat team members has had a previous exposure. How may this affect the operation if this individual becomes exposed?

4. List the terminology that you will look for once the exact chemicals are known.

5. What general chemical classification could these phase one chemicals come from?

6. Discuss the differences between the acute, subchronic, and chronic testing procedures.

7. How does the curve of the toxicological data affect the decision-making process?

8. By utilizing the toxicological worksheet provided at the end of this chapter, what are the considerations?

Body Systems and the Environment

Objectives

Given a hazardous materials incident you should be able to recognize and identify the routes of entry of a chemical and predict the possible injury incurred by the patient based on the properties of the material. You should also possess an in-depth knowledge of the pathophysiology involved with chemical exposures to the skin, respiratory system, cardiovascular system, and eyes as well as the effects of heat, cold, or smoke exposure on the body systems.

As a hazardous materials medical technician, you should be able to:

- Recognize potential hazards to medical providers, transporters, and hospital employees and take the necessary action to lessen these hazards.
- Understand the routes of entry a chemical can take to gain access into the body.

Inhalation	Ingestion
Absorption by skin and eyes	Injection

- Describe the anatomy and physiology of the skin and how injury to the skin takes place.
- Describe the anatomy and physiology of the eye and how injury to the eye takes place and how to rate the severity of that injury.
- Demonstrate the use of specialized eye medications and equipment.

  Alcaine, ponticaine, etc.

  Use of nasal cannula for irrigation.

  Use of Morgan Lens.
- Describe the anatomy and physiology of the respiratory system and how an injury to that system takes place.

- Identify specialized respiratory equipment that may be useful during a respiratory system chemical injury.

  BVM with PEEP valve.

  Pulse oximetry, uses and predictable effects from chemical exposures.

  Field ventilators.
- Describe the physiologic effects suffered by the cardiovascular system during chemical exposures.
- List the common signs and symptoms associated with and the BLS and ALS treatment for the following:

  Corrosives.

  Irritants, both respiratory and topical.

  Asphyxiants, both simple and chemical.
- Identify the signs and symptoms related to heat exposure as it pertains to encapsulated entry team members.
- Identify the physiology associated with an exposure to hot, high humidity environment.
- Identify the physiology associated with acute exposure to cryogenic materials and the treatment for this injury.
- List the common chemicals produced during the combustion process and identify what fuels produce them.

CONTENT QUESTIONS

1. The four routes of entry into the body possible on hazardous materials scenes are
   A. ____________________
   B. ____________________
   C. ____________________
   D. ____________________

2. Effects experienced shortly after exposure are called ____________________ effects.
   A. chronic B. subacute C. acute D. immediate

3. The majority of work related exposures and injuries from chemicals are through what route?
   A. Skin B. Respiratory C. Injection D. Ingestion

4. The skin makes up approximately ____________________ square inches.
   A. 1000 B. 2000 C. 3000 D. 4000

5. Which layer is found on the surface of the epidermis and is meant to protect the underlying tissue?
   A. Stratum corneum
   B. Stratum granulosum
   C. Stratum spinosum
   D. Stratum basale
   E. Stratum lucidum

6. This skin injury is usually limited to localized redness, minor swelling, and occasional blistering.
   A. Burn
   B. Irritation
   C. Sensitivity
   D. Allergic reaction

7. Most chemical burns to the skin are from
   A. solvents.
   B. acids and alkalis.
   C. phosphorus.
   D. alkali metals.

8. Coagulation of the proteins found in the skin is usually the result of exposure to what type of chemical?
   A. Alkalis
   B. Acids
   C. Solvents
   D. Corrosive metals

9. Liquification of the tissues occurs during an exposure to
   A. acids.
   B. alkalis.
   C. lead.
   D. cyanide.

10. This phosphorus is dangerous because it can autoignite once in contact with the skin.
    A. White phosphorus
    B. Red phosphorus
    C. Black phosphorus
    D. Brown phosphorus

11. Contact with these metals not only cause a heat burn, but also a corrosive burn, and while reacting, can ignite. These metals should not be washed off with water because this action will cause the reaction to increase in intensity.
 A. Porous metals
 B. Liquid metals
 C. Alkali metals
 D. Ignitable metals

12. That portion of the eye globe exposed to the outside environment is about _____ of the surface area.
 A. 1/2 B. 1/4 C. 1/6 D. 1/3

13. The surface of the globe is covered with a layer of tissue called the _____ tissue.
 A. Epithelial B. Connective C. Fibrous D. Clear

14. The tear film is a combination of fluids containing
 A. water, salt, and solvent.
 B. mucin and lipids.
 C. vitreous and lipids.
 D. conjunctiva and mucin.

15. The cornea has three layers to be concerned with during the exposure to a chemical. These three layers are (choose the three) ____________________, ____________________, ____________________.
 1. endothelium
 2. sclera
 3. stroma
 4. epithelium

16. Perilimbal circulation is
 A. circulation of the tear film over the anterior surface of the eye.
 B. circulation of the blood to the globe and associated tissues.
 C. movement of the vitreous fluid in the globe.
 D. movement of tears from the eyes to the nasal passage.

Use one of the terms below to answer questions 17–19.

A. Mild burns B. Moderate burns C. Severe burns

17. Characterized by discoloration of the stromal membrane masking the detail found in the iris. No evidence of circulation within the scelera. ____________________

18. Loss of epithelial tissue but good circulation and clear cornea noted. ____________

19. Haziness of cornea noted, epithelial tissue sloughing off but good circulation noted. ____________________

20. Which chemical injury is the most devastating to the eye?
 A. Acid B. Alkali C. Solvent D. Lacrimators

21. The symptoms of stinging pain, increased tearing, and blepharospasm, with a history of being sprayed from a canaster by police attempting to control a fight is usually what type of chemical?
 A. Acid B. Alkali C. Solvent D. Lacrimator

22. The initial treatment given to a person who reports a chemical eye injury in a safe scene is
    - A. to force the eyes open to determine the extent of injury.
    - B. to place an irrigation lens in the eye and begin flushing with water.
    - C. to start an IV and give pain medication.
    - D. to irrigate the eyes with the most accessible bland solution.

23. Using an irrigation lens such as the Morgan should be the first line of treatment for a victim of chemical eye exposure.
    - A. True
    - B. False

24. For irrigation to be accomplished, the exposed eye(s) need not be held open because irrigation over closed eyes still accomplishes good irrigation.
    - A. True
    - B. False

25. Lung capacity in a healthy adult is approximately
    - A. 2000 ml
    - B. 3000 ml
    - C. 5000 ml
    - D. 6000 ml

26. Passageways from the mouth to the alveoli not only provide a route for air to travel but also
    - A. moisten or dry the air.
    - B. filter the air.
    - C. warm or cool the air.
    - D. all of the above.

27. The normal adult lung contains approximately ____________________ alveoli and has a surface area of ____________________ square feet.
    - A. 12 million; 500
    - B. 100 million; 300
    - C. 5 million; 2000
    - D. 300 million; 1000

28. Molecular dispersion of a solid or a liquid in the air is called (a, an)
    - A. vapor.
    - B. gas.
    - C. mist.
    - D. aerosol.

29. The solubility of an inhaled chemical determines where in the respiratory system the initial injury will develop. Therefore non water soluble irritants will usually injure the
    - A. mouth.
    - B. trachea.
    - C. mainstem bronchi.
    - D. alveoli

30. An example of a simple asphyxiant is
    - A. carbon monoxide.
    - B. carbon dioxide.
    - C. hydrogen sulfide.
    - D. hydrogen cyanide.

31. Chemical asphyxiants are different than simple asphyxiants in that they
    - A. interfere with the transportation or use of oxygen in the body.
    - B. interfere with the alveoli's ability to transfer oxygen.
    - C. displace all of the oxygen in the lungs.
    - D. cause the cells of the fine bronchioles and alveoli to leak.

32. Chemically induced pulmonary edema results from
    - A. a high pulmonary pressure.
    - B. a heart weakened due to chemical exposure.
    - C. injured cells at the distal end of the lower respiratory system.
    - D. extensive secretion production.

33. In severe chemically induced pulmonary edema, the lungs can hold up to ______________________ of fluid.

A. 1000 ml B. 500 ml C. 2 liters D. 4 liters

34. Pulse oximetry measures

A. the amount of oxygen in the blood solution.
B. the amount of oxygen carrying potential of the blood.
C. the amount of oxygen bound with hemoglobin.
D. the amount of carboxyhemoglobin.

35. The oximeter can give unusually high readings with all of the following chemical poisonings ***except***

A. carbon monoxide.
B. cyanide.
C. hydrogen sulfide.
D. nitrate.

36. Define the meaning of the acronym PEEP.

____

____

37. Define the meaning of the acronym CPAP.

____

____

38. What is the difference between the two described in 36 and 37.

____

____

39. Hyperbaric oxygen can be used for some patients experiencing symptoms of a hazardous materials exposure. Which are the two recognized uses involving hazardous materials?

____

____

40. The loss of fluid from sweating and other insensible loss can reach as high as

A. 500 cc an hour.
B. 1000 cc an hour.
C. 2000 cc an hour.
D. 3000 cc an hour.

41. Water is absorbed an average of __________ faster than power drinks.

A. 5–10%
B. 10–23%
C. 25–34%
D. 35–39%

42. Describe the difference between regular hydration and forced hydration.

____

____

____

____

43. A loss of 5% body weight of a 200-pound worker while functioning in an encapsulated suit would require a theoretical rehab time of ____________________ before full recovery.

44. The highest percentage of heat loss occurs from which method?

A. Convection
B. Conduction
C. Evaporation
D. Radiation

45. Heat stroke has an estimated mortality of

A. 5–10%. B. 15–25%. C. 40–45%. D. 50–80%.

46. Name and describe three types of cold injuries.

____

____

____

____

____

____

47. What type of injury would you expect to see if a truck driver off loading a cryogenic experienced a sudden leak that he attempted to stop by placing his hand over the product?

____

____

____

____

____

____

CONTENT APPLICATION #1

You have just arrived on the scene of an auto accident involving a van and an auto. The van apparently broadsided the smaller vehicle causing minor injury to the driver of the smaller car. The van suffered significant damage to the front end, ejecting the windshield and lifting the hood which exposed the engine. Upon examination of the van driver, you find a 35-year-old male holding his hands over his eyes. When you are able to move his hands, you find that his eyes are completely white (including the corneas) and some tissue is sloughing from the upper and lower lids. The distinct odor of acid fumes is noted and a quick look at the front of the auto discloses a crushed battery. You are assigned to treat this victim. What are the priorities after your initial assessment? How would you treat the chemical burns?

CONTENT APPLICATION #2

As a hazmat paramedic you respond with the team to an ammonia leak in Smith's Blueprinting and Graphics Service. Upon arrival your unit is met with several employees whose eyes are watering and who are coughing. They direct you to the office manager who attempted to fix the leaking hose in the office. The other employees stated that they left as soon as the odor was noted, but the manager returned inside for an additional 5 to 10 minutes in an attempt to fix the leak. He stumbled out of the front door and fell to the ground. He now presents with the distinct odor of ammonia. His skin is diaphoretic, pale in color, and cool to the touch. His breathing is evident due to the loud rattle coming from his upper respiratory system. Auscultation with a stethoscope finds rales in the lower fields and wheezing and rhonci in the upper airways. His pulse is rapid at 138 and BP of 144/94. A pulse oximeter indicates an SaO_2 of 78%. What would you expect to be the physiologic problem of the patient? Explain in detail. What steps would you take to treat the patient?

CHAPTER 5 SCENARIO

You are the paramedic on ambulance 5, an ALS unit that specializes in hazardous materials emergencies. You have been dispatched in conjunction with two other ALS units to assist a fire unit already on the scene of a small fire where several firefighters are reported down. Upon your arrival, the incident commander informs you that the fire involved a residential garage. The fire was knocked down in about 15 minutes with overhaul taking an additional 30 minutes. The initial complaints by the firefighters were brushed off as complaints due to the hot summer day. The first firefighters on the scene complained of burning eyes as they approached the house before they put on their breathing devices. Once the fire was knocked down and overhaul was under way, several firefighters removed their masks to complete the overhaul. Two of them removed their jackets because of the 98° temperature and a reported humidity of 85%.

A quick visual assessment indicates that four firefighters are injured, all having shortness of breath, a productive cough, and complaining of burning in the chest. Two of the firefighters are cyanotic, have a decreased level of consciousness, and are displaying involuntary muscle jerks in both their legs and arms. All have reddened eyes and are tearing profusely.

Scenario Questions

1. Is decontamination of the patients immediately necessary? Why or why not?

____

____

____

____

____

2. If you have decided to decontaminate, what areas of the body would be targeted for extensive decontamination?

____

____

____

____

____

3. What routes of entry has the chemical probably taken?

____

____

____

____

____

4. Would you consider this a systemic poisoning?

5. Would it be more important to identify the chemical affecting the firefighters or to treat their symptoms?

6. Could heat be a factor in the absorption of the chemical?

7. Could heat be a factor in the symptoms displayed?

8. When a full assessment is possible, what types of data should be gathered on these patients?

9. What specialized equipment may be needed for the treatment of these patients?

10. Do any special arrangements have to be made with the hospital prior to transport?

Treatment Modalities

Objectives

Given a hazardous materials incident, the ALS provider should have the competencies necessary to provide advanced life support to persons injured by chemical exposures. Students completing this section are expected to recognize the pathophysiology of individual poisonings and to determine the cause for signs and symptoms displayed by the patient.

As a hazardous materials medical technician, you should be able to:

- Identify the routes of exposure, means of decontamination, pathophysiology involved in the exposure and resulting injury, signs and symptoms expected, basic and advanced life support for the injury, and the common places that these materials are found.

  Carbon monoxide

  Cyanide and cyanide related materials

  Hydrogen sulfide

  Hydrofluoric acid and hydrogen fluoride gas

  Nitrites and nitrates

  Phenol and phenol containing products

  Organophosphates and carbamate insecticides

  Antipersonnel chemical agents

  Radiation and radioactive materials

CONTENT QUESTIONS

1. Carbon monoxide has
   A. the odor of bitter almonds.
   B. the odor of rotten eggs.
   C. no odor but irritates nose and throat.
   D. no discernible odor.

2. Carbon monoxide is
   A. an irritant.
   B. a simple asphyxiant.
   C. a chemical asphyxiant.
   D. a corrosive.

3. Symptoms from carbon monoxide poisoning are usually initially present after a carboxyhemoglobin level of greater than __________ is attained.
   A. 10% B. 25% C. 50% D. 80%

4. The expected oximetry reading from a patient poisoned with CO would be
   A. low. B. high. C. no change from normal. D. unable to be read.

Match the expected half life of carboxyhemoglobin with the treatment.

5. COHg T1/2 of 6 hours ______ A. hyperbaric oxygen
6. COHg T1/2 of less than 1 hour ______ B. 100% oxygen
7. COHg T1/2 of 1.5 hours ______ C. room air

8. Hydrogen cyanide has
   A. the odor of rotten eggs.
   B. the odor of bitter almonds.
   C. the odor of cat urine.
   D. acid irritation but no odor.

9. Cyanide is
   A. a chemical asphyxiant.
   B. a simple asphyxiant.
   C. a respiratory irritant.
   D. none of the above.

10. Cyanide's signs and symptoms are related to its interference with
    A. the oxygen carrying capability of the blood.
    B. the use of oxygen in the cell.
    C. the intake of oxygen into the body from the outside environment.
    D. absorption of fluids into the GI system.

11. Cyanide is said to have a characteristic odor of
    A. sweet onions.
    B. rotten eggs.
    C. bitter almonds.
    D. sewer gas.

12. Why can only some individuals smell the characteristic odor associated with cyanide?

    ____

    ____

13. Some cyanide is found in the everyday environment such as cigarette smoke and food. How is the body able to deal with these small amounts?

____

____

____

____

14. Of the listed respiratory signs of cyanide poisoning, which are early signs and which are considered late signs? Signify by marking E for early and L for late.
    - **A.** decreased respiratory rate ________
    - **B.** dyspnea ________
    - **C.** tachypnea ________
    - **D.** respiratory depression ________
    - **E.** apnea ________
    - **F.** hyperpnea ________

15. Of the listed cardiovascular signs of cyanide poisoning, which are early signs and which are considered late signs? Signify by marking E for early and L for late.
    - **A.** hypotension ________
    - **B.** reflex bradycardia ________
    - **C.** flushing ________
    - **D.** hypertension ________
    - **E.** acidosis ________
    - **F.** tachycardia ________
    - **G.** ST changes ________
    - **H.** AV nodal or intraventricular rhythms ________

16. List 5 industrial settings and 2 types of pesticides where cyanide might be found.

____

____

____

____

____

17. What are the three drugs used to treat cyanide poisoning? Give their effects and list the dosage of each. Describe why they are given in the prescribed order.

____

____

____

18. What treatments other than drugs may be useful for cyanide poisoning?

____

____

19. List three properties of hydrogen sulfide.

____

____

____

20. "Gas eye" or "Sewer eye" is terminology used to describe what?

____

____

21. Hydrogen sulfide is an irritant. What part of the body does it irritate?

A. Skin
B. Eyes
C. Respiratory system
D. Mucous membranes
E. All but a

22. Signs and symptoms from HS poisoning include all but

A. bronchospasms.
B. chemically induced pulmonary edema.
C. euphoria
D. headache.

23. List 3 places that hydrogen sulfide may be found.

____

____

____

24. Is decontamination of a patient who has been removed from a hydrogen sulfide environment necessary? Explain.

____

____

____

25. List the medications and dosages used to treat hydrogen sulfide poisoning. Is it necessary to give these in a particular order? Explain.

____

____

____

____

____

26. Are there any specific hospital treatments used to treat the victim of hydrogen sulfide poisoning? If so, why are they selected?

____

____

____

27. List the properties of hydrofluoric acid.

____

____

____

____

____

28. Identify the two specific injuries caused from contact with hydrofluoric acid.

A. Caustic skin burn
B. Increased urination
C. Decalcification of the bone
D. Increased cardiac output

29. Hydrofluoric acid seeks out and binds with what two elements?

A. Potassium
B. Magnesium
C. Calcium
D. Aluminum

30. Describe why the victim of a hydrofluoric exposure complains of excruciating pain at the site of contact.

____

____

____

____

31. List 4 industries where hydrofluoric acid is commonly used.

____

____

____

____

32. Identify the treatment used if a victim splashes hydrofluoric acid in his eyes.

____

____

____

____

33. If hydrofluoric acid makes contact with the skin, what is the recognized treatment? Are there other mixtures that can be used if the specific antidote is not available?

____

____

____

____

____

34. If a victim inhales the fumes of hydrofluoric acid and complains of increased pain and burning, what is the acceptable treatment?

____

____

____

____

35. Nitrate or nitrite poisoning is many times caused by someone using chemicals containing one of these compounds for recreational reasons. Knowing the physiology of the injury, what kind of "high" can a person expect to experience?

____

____

____

____

36. Identify all of the routes of entry that may be associated with nitrate/nitrite poisoning.

A. Inhalation **B.** Absorption **C.** Ingestion **D.** Injection

37. Nitrates and nitrites change the valence of iron (Fe) in the hemoglobin from ferrous (Fe^{++}) into ferric (Fe^{+++}) iron changing hemoglobin into ________________.

38. The signs and symptoms associated with nitrite or nitrate poisoning include all but

A. decreased use of oxygen on the cellular level.

B. throbbing headache.

C. dizziness.

D. tachycardia.

E. flushing of the neck and face.

39. Where are products containing nitrates and nitrites found in industry? List 6.

____

____

____

____

____

____

40. What is the antidote for nitrite poisoning? What physiologic change takes place that allows the antidote to work?

____

____

____

____

____

41. What other treatments should be undertaken in conjunction with the antidote administration?

____

____

____

42. Phenol was initially used in a hospital setting

A. as a cleaning solution.
B. as a disinfectant.
C. as a deodorizer.
D. as a cough medicine.

43. Phenol toxicity is displayed as

A. CNS stimulation.
B. CNS depression.
C. tachycardia.
D. rapid breathing.

44. Typical symptoms include all but

A. hypothermia.
B. tachypnea.
C. cardiac depression.
D. loss of vascular tone.

45. Treatment of phenol toxicity begins with proper decontamination. List the agents used to properly decontaminate a victim of phenol exposure.

____

____

____

46. List 10 uses of phenol and areas where it may be found.

47. When responding to an adult who accidentally ate a small portion of rat poison which is a known anticoagulant, which anticoagulants would you be the most concerned about?

48. If your patient had ingested a herbicide, of the three, paraquat, diquat, and morfamquat, which would you be the most concerned about? Why?

49. Why are chlorinated hydrocarbons such as DDT no longer used in the United States as a pesticide?

50. Of the following list, which organophosphate is the most toxic?

A. Parathion **B.** Malithion **C.** Diazinon **D.** Sarin

51. What makes some organophosphates more toxic than others? Explain.

____

____

____

52. Explain the physiology of an organophosphate poisoning. What causes the patient to salivate so much?

____

____

____

____

53. Several acronyms were listed in the text. List the correct words represented by the acronyms:

A. SLUD ____

B. SLUDGE ____

C. DUMBELS ____

54. List the symptoms of organophosphate poisoning displayed by each of the following systems:

A. CNS

____

____

B. Somatic

____

____

C. Sympathetic

____

____

D. Parasympathetic

____

____

55. Is it necessary to decontaminate patients contaminated with an organophosphate? If so, what solutions are appropriate?

____

____

____

____

56. What is the physiology related to the injury caused from organophosphates?

57. If it is true that organophosphates are only active in your body for one to two days, why are the effects sometimes felt for up to 30 days?

58. What does the antidote atropine do in the body that alleviates the problems associated with organophosphate poisoning?

59. What dosages of atropine should be used? How is the dosage determined?

60. Protopam chloride is also used in the treatment process. What is the advantage to using this drug?

61. Carbamate insecticide poisoning is similar to organophosphate poisoning except for one major difference. What is the major difference?

62. Is protopam used for carbamate poisoning? Why or why not?

63. Of the antipersonnel chemical agents, OC (oleoresin capsicum) is the most common. How does it affect the person it is sprayed on?

____

____

____

64. When OC is sprayed on a person, are the effects tissue damage? If not, how can it be treated?

____

____

____

65. Ionized radiation is found in several ways. What are they and define each.

____

____

____

____

____

66. What are the principles of protection?

A. Time, distance, shielding, and quality
B. Time, blocking, shielding, and quality
C. Time, distance, shielding, and quantity
D. Time, blocking, shelving, and quality

67. What is meant by time being a principle of protection?

____

____

68. Name four of the common sites for radiation storage or usage.

____

____

____

____

69. Define each of the 4 types of injuries.

A. External irradiation

____

____

B. Contamination

C. Incorporation

D. Irradiation

CONTENT APPLICATION #1

The call is received on a December morning at 2:00 A.M. The temperature is 34° and it has been raining most of the evening. You are dispatched to a person complaining of shortness of breath, headache, and dizziness. Upon entering the residence you notice a small dead kitten just inside the door. The person is found in the bedroom where there is evidence of vomit on the bed sheets. The patient is a 30-year-old female who appears pregnant. She is slow to answer your questions and when she does her speech is slurred. A pulse oximetry indicates 99% with a pulse of 122. The blood pressure is 160/88. After applying oxygen and assessing the patient, your partner tells you that he is now experiencing a headache and dizziness. What action would be appropriate at this point? What do you think the problem could be? Is this patient in a serious condition that needs immediate specialized attention? Explain.

CONTENT APPLICATION #2

You respond to a man in a potato chip manufacturing business who is down. It is a midsummer day with temperatures reaching 90 degrees. Upon arrival, you are directed to an area where the potatoes are initially unloaded and washed. The patient was removed from the inside of a large holding area where the potatoes are washed and the contaminants are filtered from the wash water. Apparently, one of the filters became clogged with dirt and organic material and the patient was sent to clean them out. When he did not return in a reasonable period (about an hour), other workers went looking for him and upon finding him unconscious, dragged him outside into the fresh air. The patient is breathing rapidly at 36 times a minute and his blood pressure is 180/100 with a pulse rate of 62. His skin color is flushed, cold to the touch (he is soaking wet), and the pulse oximeter indicates 100%. His eyelids are also very red and appear swollen. He does not respond to your commands but does react to painful stimuli. His clothing is wet, dirty and smells of rotten potatoes. What would you expect this condition to be related to? How would you treat a patient who presents with this history and physical findings? Explain the physiology of such an injury.

____

____

____

____

____

____

____

____

____

____

____

____

____

CHAPTER 6 SCENARIO

You are a member of an ambulance responding to a chemical spill in a semiconductor manufacturer. Dispatch reports that there are 4 patients that have been removed from the structure by the hazmat entry team. Upon your arrival you notice that the zones have been set up and the last patient is just entering the cold zone from the decontamination corridor. All four patients are unconscious. The hazmat officer reports that a forklift inside the building collided with some shelving causing it to fall. The shelving stored several different chemicals. You are handed a stack of MSDS sheets that were provided by the manufacturing facility. Among the chemicals you notice a variety of acids (including 70% hydrofluoric, 10% sulfuric, and 20% acetic acid), 50% sodium hydroxide, formonitrile, and a solvent, methylene chloride.

Patient #1 was the driver of the forklift. She is a 35-year-old female who is unconscious but responds to painful stimuli. Further signs include flushed skin, BP of 180/100, pulse of 54, and the EKG indicates a first degree block with sinus bradycardia, pulse oximetry is 100%, and she is hyperventilating at 36 times per minute. There are no signs of trauma but a garlicky, almond smell has been reported by the rescuers.

Patient #2 is a 28-year-old man who has a 6-centimeter laceration on the forehead and several open wounds on the legs. Chemical burns are noted to both legs from the knees to the ankles. The entry team stated that he was removed from under the debris that fell from the shelves. The wounds appear milky in color and the skin is slick to the touch with blistering noted at the outer edge. The skin's general appearance other than the burn is normal in color and temperature. The BP is 130/84, a pulse of 60, SaO_2 reading of 95%, and respirations of 32. An EKG indicates a prolonged QT interval.

Patient #3 is a 60-year-old man who was the supervisor in the shop where the accident happened. He was the first to respond and while he was pulling boxes off of patient #2 and attempting to drag him free of the debris, he clutched his chest and dropped to the ground. His skin is pale, cool, and diaphoretic. His BP is 90/40 with an irregular pulse rate of 40. His breathing is shallow at 12 times a minute. An EKG indicates atrial fibrillation with frequent multifocal PVCs. Pulse oximetry indicates an oxygen saturation of 88%.

Patient #4 is an 18-year-old female. She was working in the front office when she heard the crash. She also responded to help. She attempted to move a 10 gallon container of liquid and upon lifting it from the leaning shelf the container failed spilling the fluid over her head, face, and down her chest, soaking her clothing. She worked removing debris until she eventually lost consciousness. She presents with flushed skin, rapid breathing at 36 times a minute, pulse of 140, and a BP of 160/100. The pulse oximeter reads 100%. She still has a faint odor of solvent in her hair. An EKG indicates ST depression with occasional PVCs.

Scenario Questions

1. Evaluating the symptoms presented by patient #1, what do you expect to be the primary offending chemical?

____

2. What are some associated symptoms that could verify your diagnosis?

____

____

3. You have confirmed your diagnosis. Now how would you treat the patient? What supportive care would you use (include any drugs and dosages)?

____

____

____

____

____

4. Evaluating the symptoms presented by patient #2, what do you expect the offending chemical to be?

___

5. Is an exposure of this amount of body surface area significant? Why?

___

___

6. You have confirmed your diagnosis. Now how would you treat the patient? Include the supportive care, drugs and dosages.

___

___

___

___

7. Is it possible that patient #2 is also suffering from the effects of multitrauma? If so, explain.

___

___

___

___

___

8. Evaluating the signs and symptoms presented by patient #3, what do you expect the exposure to be?

___

___

9. Is it possible that patient #3 is also suffering from a medical condition? If so, explain.

___

___

___

___

___

10. Could an exposure contribute to the exacerbation of a preexisting medical condition?

___

11. Evaluating the signs and symptoms of patient #4, what do you expect the offending chemical to be?

12. Does patient #4 need further decontamination? Explain.

13. You have confirmed your diagnosis. Now, how would you treat the patient? Include any specialized equipment, drugs and dosages, if appropriate.

14. If this situation happened today, is your unit, agency, or district prepared to handle it?

15. What would you suggest is necessary to bring your agency up to the standards needed for emergency medical hazardous materials response?

Medical Surveillance

Objectives

Before, during, and after the hazardous materials event, with the medical director you should formulate the level of surveillance that should occur on as many considerations presented. It is up to the local agency's management in association with the resources personnel to meet on a regular basis in order to evaluate local resources. You should be able to organize the components of a medical surveillance program given the resources available within the local response area.

When presented with a hazardous materials incident, you should be able to identify the capabilities of the medical sector in terms of medical surveillance that may be provided. In addition to the medical pre-planning that should occur and the long term considerations of medical surveillance you should compare signs and symptoms with hazardous materials scene variables.

As a hazardous materials medical technician, you should be able to:

- List the pertinent information for analyzing the systems capabilities for:
  Preincident medical surveillance.
  On-scene medical surveillance.
  Follow-up medical surveillance.
- Identify the foregoing components and how to manage such a program within the local system, utilizing available resources.
- Describe the importance of the following and the surrounding considerations of each:
  Preemployment physical
  Annual physical
  Cursory physical
  Follow-up physical
- Discuss the federal regulations that identify a medical surveillance program.
- Describe the individual components in detail, of the preincident medical surveillance program.

Chest x-rays
Pulmonary function test
Pulmonary diffusion test
Vision test
Auditory test
ECG and stress ECG
Blood work
 Blood chemistry
 Complete blood count
 Biological monitoring of known exposure
Urine analysis
Conventional physical exam

- Discuss the need for interval medical histories.
- Describe the individual components in detail, of the cursory medical exam:

Blood pressure
Pulse
Respirations
Pulse oximetry
Lung field auscultation
Other considerations
ECG 12 lead
Tympanic temperature
Weight
Forced hydration
General sensorium
Blood drawing

- Describe the exit physical in terms of the above components. Discussing in detail each area of concern.
- Discuss the pros and cons of field blood draw analysis.
- Discuss the use of exclusion criteria.
- Identify the pros and cons of medical status form in terms of the law.
- Identify the components of the Program Review segment within the medical surveillance program.
- Identify the need for critical incident stress debriefing within the context of medical surveillance.
- Describe how you as a student may establish a medical surveillance program within your system utilizing existing resources while maintaining cost constraints.

CONTENT QUESTIONS

1. The system that is designed to maximize the health of the emergency worker, while at the same time minimizing the health risks is called
   - A. Ryan White Act.
   - B. 29 CFR 1910.120.
   - C. medical surveillance.
   - D. medical check-up.

2. The program that revolves around work safety, involving the physical controls that we can employ to reduce the possibility of an injury is called
   - A. medical surveillance.
   - B. physical controls.
   - C. engineering controls.
   - D. administrative controls.

3. The second half of the program revolves around what types of controls?
   - A. Medical surveillance
   - B. Physical controls
   - C. Engineering controls
   - D. Administrative controls

4. Match the function with the segment of medical surveillance.

A. Personnel protective gear _____	1. Engineering controls
B. Consensus standards _____	2. Administrative controls
C. Regulations _____	
D. Outline of approaches that are used prior to, during, and after the event _____	
E. Medical monitoring of the entry and decon team _____	
F. Decontamination techniques _____	
G. Standard operating guidelines _____	

5. Describe four reasons why we need to monitor one's health status prior, during, and after the hazardous materials incident.

   A. ____

   ____

   ____

   B. ____

   ____

   ____

   C. ____

   ____

   ____

   D. ____

   ____

   ____

6. OSHA advocates that a program that evaluates one's health be maintained by the employee.

   A. True B. False

7. The exact content of a medical surveillance program is outlined in 29 CFR 1910.120 paragraph q.

   A. True B. False

8. 29 CFR 1910.134 identifies the need for appropriate training for those individuals who use a respirator in the course of their assigned work.

   A. True B. False

9. Identify the following statements regarding OSHA recommendations of medical surveillance as true or false. Mark T, for true, F for false.

   A. Any firefighter who, in the normal course of duty, is not on a hazardous materials team, must be a part of the medical surveillance. ____________

   B. Any emergency worker, if exposed to any hazardous material for 20 days or more within one year (the levels must be equal or above the OSHA standard for exposure), or there is an injury due to exposure. ____________

   C. All members of the hazardous materials team. ____________

   D. The frequency of medical tests is limited under the responsibilities of the examining paramedic. ____________

   E. The examining physician has the right to go beyond the federal or state suggestions when in his/her opinion the worker's involvement is such that a health hazard exists. ____________

   F. Within areas of concern, air monitoring shall be employed and documented. ____________

   G. The employee is responsible for the cost of the medical surveillance program. ____________

   H. The employer is required to establish and maintain the medical surveillance program for all eligible employees. This may, under certain conditions, include any employee who, during the work environment, may be exposed to half the OSHA exposure limit. ____________

   I. Frequency of the physical shall be every 12 months. ____________

   J. Upon retirement or termination from the area of employment, a physical exam shall take place. In conjunction with the medical records, a job description of the employee should be maintained within the records to identify the working responsibilities of said employee. ____________

10. Documentation of the whole program plays an important part in medical surveillance. Under the ______________________, employers are required to maintain health related records during employment and ______________________ years thereafter.

    A. State law; 7 C. State and local law; 30

    B. Federal Register; 7 D. Federal Register; 30

11. In 1990 the Americans with Disabilities Act (ADA) was passed. This act reminds the employer
    - A. of responsibilities they have toward all employees.
    - B. of the expanded laws surrounding individuals with disabilities.
    - C. that discrimination against anyone who can perform the essential job function is prohibited.
    - D. of all the above.

12. The ADA states that preemployment medical physicals
    - A. are prohibited as an exclusion from employment.
    - B. are allowed only if the employer has given a conditional offer of employment.
    - C. conditional offer of employment is contingent on the outcome of the exam.
    - D. all the above apply.

13. As a consideration during the physical assessment, a brief medical history of off-duty employment is important.
    - A. True
    - B. False

14. A health assessment should be well rounded in order to establish a good medical "view" of the employee. This assessment, although suggested for hazardous materials team members, is not required for all emergency responders. Describe the content of such a medical.

    ____

    ____

    ____

    ____

15. Describe the use of X-rays and the pulmonary function test and explain why this is assessed.

    ____

    ____

    ____

    ____

16. What is a pulmonary diffusion test?

    ____

    ____

    ____

    ____

17. How are the above three tests utilized within medical surveillance?

18. What vision tests should be performed?

19. What must occur to preserve one's hearing?

20. Describe the type of blood chemistry that should be incorporated within a medical surveillance program.

21. Describe in detail the annual physical exam.

22. Hazardous materials team members and firefighters who have been exposed to hazardous substances must have a level of medical surveillance applied to them.

A. True **B.** False

23. Interval history is a questionnaire used to identify new medical concerns, exposures, and/or illnesses.

A. True B. False

24. Interval exams and interval histories are one and the same.

A. True B. False

25. Annual physicals are a necessity for all individuals involved with hazardous materials emergency response.

A. True B. False

26. Describe the difference between biological monitoring and the annual physical.

____

____

____

____

____

27. Describe in detail the components of the cursory physical exam.

____

____

____

____

____

28. Describe three important criteria that relate to the cursory physical:

1. _______________

____

____

____

2. _______________

____

____

____

3. _______________

____

____

____

29. Through the exam process, we can test and judge the individual as to his/her physical preparedness.

A. True **B.** False

30. The on scene cursory medical exam consists of two segments. What are they?

A. Medical history and the entrance exam
B. Interval history and the exit exam
C. Entrance exam and the interval exam
D. Entrance exam and the exit exam

31. A common medical problem of team members at the scene of a hazardous materials incident is

A. heat stroke.
B. heat exhaustion.
C. A only.
D. B only.

32. How can one prepare for the above medical problem?

A. Through forced hydration
B. Through heat acclimation
C. Through medical monitoring
D. Both A and B.

33. The percentage of decrease in body weight considered to be non-life threatening is

A. 5%. **B.** 3%. **C.** 4%. **D.** 6%.

34. While on scene, specific medical history may include what questions?

____
____
____
____
____

35. This cursory exam should include, but is not limited to:

____
____
____
____

36. What is the reason for exclusion criteria?

____
____
____
____

37. Describe what is considered on the entrance physical relating to blood pressure.

38. Describe what is considered on the entrance physical relating to the pulse.

39. Describe what is considered on the entrance physical relating to respiratory status.

40. Describe what is considered on the entrance physical relating to pulse oximetry.

41. Describe what is considered on the entrance physical relating to lung auscultation.

42. Describe what is considered on the entrance physical relating to an EKG.

43. Describe what is considered on the entrance physical relating to temperature.

44. Describe what is considered on the entrance physical relating to weight.

45. Describe what is considered on the entrance physical relating to general sensorium.

46. Describe the pros and cons of drawing blood at the scene of a hazardous materials event.

47. Under what four conditions shall a full cursory medical be performed?

48. Describe the legality of the medical status sheet.

____

____

____

____

49. The exit physical is part of the termination of the incident. It can also guide the command staff toward:

A. Resources required to complete the incident
B. Future personnel needs
C. A and B
D. None of the above

50. Exit physicals should be, at a minimum, within what percent of entrance vitals?

A. 5% B. 10% C. 15% D. 20%

51. Who should receive a cursory medical?

A. Entry team members
B. Decontamination team
C. Medical treatment team
D. All the above

52. Describe the basic exclusion criteria.

____

____

____

____

53. If exposure occurs, a set of follow-up physicals is indicated. How often should these occur?

A. Every three months
B. Every nine months
C. As often as deemed necessary by the medical director
D. A and C

54. The process of peer support and specifically trained mental health providers is called:

A. Medical surveillance
B. Cursory medical
C. Critical incident stress debriefing
D. Engineering controls

55. If the incident warrants a complete debriefing, the debriefing process should occur within __________ hours of the incident.

A. 12 B. 24 C. 48 D. 72

56. Identify the seven components of a debriefing:
   1. ______
   2. ______
   3. ______
   4. ______
   5. ______
   6. ______
   7. ______

57. How many days after the debriefing should pass before an operational critique can take place?
   A. 1 day B. 2 days C. 3 days D. 5 days

58. Describe the circumstances that call for a critical incident stress debriefing.

______

______

______

______

CONTENT APPLICATION

By utilizing the information you have gained from this chapter, how would you organize the scene medical surveillance program and what would your considerations be regarding the following?

1. Placement of treatment sector in relation to medical sector

   ______

   ______

2. Transporation and staging area

   ______

   ______

3. Complement of personnel

   ______

   ______

4. Additional resources

   ______

   ______

CHAPTER 7 SCENARIO

Your department has allocated a large portion of the hazardous materials response budget to the establishment of a hazardous materials medical surveillance program. The medical director and yourself have just returned from a class describing the fundamentals of such a program. The medical director is very concerned and has persuaded the chief of your department to allocate initial start-up and continued financial support. Both the chief and the medical director are concerned about the legal ramifications of such a program and the continued support structure.

You have been placed in charge of the development, implementation, and maintenance of such a program. In order to assist in your project you elicit the help of a mutual aid jurisdiction. They have just started to develop a similar program, but have been hampered by the lack of financial support from their department. It is your goal to develop, implement, and educate the response personnel in both jurisdictions, while implementing a cost effective program.

From your notes that were taken at the medical conference you have just returned from you start to develop your program:

Scenario Questions

1. How should prehire physicals be handled and what ramifications do the existing programs have?

____

____

____

____

____

2. Who could provide the annual physical and what are the components of such an exam?

____

____

____

____

____

3. What are the components of the cursory exam and how will this information be routed to the appropriate medical authority?

____

____

____

____

____

4. Considerations of the cursory exam:

   A. Elements of the exam are?

   ____

   B. Should blood drawing, blood alcohol, or blood sugar be a part of such an exam?

   ____

   C. What are the exclusion criteria that should be established?

   ____

   D. If a team member becomes exposed, what are the procedures?

   ____

5. How will you handle the review of such a program?

   ____

   ____

   ____

   ____

   ____

6. If there is a need for a critical incident stress debriefing, how and who will handle this part of the operation and how will they do it?

   ____

   ____

   ____

   ____

   ____

7. What other considerations will you have that may have far-reaching ramifications?

   ____

   ____

   ____

   ____

   ____

Chapter 8

Hazardous Materials Considerations for Hospitals

Objectives

At a hazardous materials incident where a patient has been injured due to an exposure and transportation to a medical facility is involved, or when a hazardous materials incident takes place within the hospital facility, the student must understand the complexities and special considerations of the effects a chemical event may have in a health care setting. Furthermore, the student should identify the importance of developing a preplan, preparing the emergency department for chemical emergencies, and the importance of regional poison control centers to hospital staff when faced with contaminated patients needing emergency medical care.

As a hazardous materials medical technician, you should be able to:

- Identify the goals that hospital receiving facilities must set for preplanning to be completed.
- Understand key points to note for identifying patients who are injured by exposure to hazardous materials.
- Give an overview of incident command when working within the hospital supervisory structure.
- Recite step-by-step the considerations to be addressed when:

  Receiving the call.

  Setting up a decontamination area.

  Directing the arrival of the patient.

  Providing decontamination.

  Selecting the proper level of personal protective equipment.
- Identify the role of the regional poison control center when a hospital is faced with an in-house hazardous materials emergency or the arrival of a contaminated patient.

CONTENT QUESTIONS

1. For hospital staff, where are the two major concerns involving hazardous materials and why are these areas becoming more dangerous?

____

____

____

____

____

2. Several steps must be taken to insure the safety of both staff and patients. Name two.

____

____

3. Resources used for training hospital staff include all but
   A. industrial personnel.
   B. local hazardous materials teams.
   C. American Red Cross.
   D. laboratory safety personnel.

4. At what level must field emergency responders be trained to provide decontamination of a patient?
   A. Awareness with additional training in decontamination
   B. Operations with additional training in decontamination
   C. Technician with additional training in decontamination
   D. Specialists

5. Name 5 classes of hazardous materials found in the hospital setting.

____

____

____

____

____

6. In-house emergency hazmat response teams' duties should include all but
   A. isolation and evacuation.
   B. controlling ventilation and drainage.
   C. plugging leaks in sterilization gas containers.
   D. controlling spills and providing absorbent.

7. Information on a chemical can be gained from
   - **A.** MSDS sheets.
   - **B.** NIOSH pocket guide.
   - **C.** poison control centers.
   - **D.** emergency department computer bank.
   - **E.** hospital library.
   - **F.** all of the above.

8. If the hospital staff suspects that a patient is brought from a hazardous materials emergency, what questions should be asked of the EMS provider?

   ____

   ____

   ____

   ____

9. Two or more patients complaining of the same symptoms from the same geographical location may give hospital staff a clue that hazardous materials may be involved. These symptoms include all but
   - **A.** breathing difficulties and/or coughing.
   - **B.** eye and nose irritation.
   - **C.** drooling.
   - **D.** sore fingers and toes.
   - **E.** headaches.

10. Walk-in contaminated patients are the biggest challenge to hospital staff. Explain.

    ____

    ____

    ____

    ____

11. Important information gained from field personnel who call the hospital advising them of a hazmat emergency with patients should include all but
    - **A.** the time of day.
    - **B.** the type or name of the chemical involved.
    - **C.** the signs and symptoms displayed by the patients.
    - **D.** what decontamination is involved.

12. What information should be given to the transporting agency that is enroute to the hospital with a hazmat patient?
    - **A.** How to decontaminate the patient in the ambulance
    - **B.** How soon to be at the hospital
    - **C.** Which route to take to the hospital
    - **D.** Where to go once they have arrived at the facility

13. A hospital decontamination area should include all but
- **A.** being inside or outside.
- **B.** dim lighting.
- **C.** warm running water.
- **D.** a means of collecting runoff.

14. If a decontamination area is chosen inside the hospital, which is not a consideration?
- **A.** Access into the decon room from outside doorway
- **B.** Ability to isolate air conditioning and heating systems
- **C.** Separate drainage systems
- **D.** Ability to control electricity into the room

15. Patient decontamination provides all but
- **A.** removing the contamination from the patient.
- **B.** lessening the chance of contamination to patient care workers.
- **C.** stopping any injury that has already occurred.
- **D.** limiting the injury to the patient.

16. What protective equipment is typically not available to emergency room staff?
- **A.** Self contained breathing apparatus
- **B.** Gloves
- **C.** Gowns
- **D.** Eye protection

17. Name the 6 requirements that must be met to become a regional poison control center.

____

____

____

____

____

____

CONTENT APPLICATION #1

You are on the scene of an overturned tanker carrying hydrochloric acid. The product covers the roadway and many autos drove through the spill prior to your arrival. Several good Samaritans stopped to assist the truck driver who was trapped under the tractor near a pool of the product. Several of the good Samaritans and the truck driver are displaying respiratory and other irritant symptoms. The victim's clothing was removed and a decontamination provided on the scene. You call the closest hospital using your department radio. What facts should you relate to the hospital? What information would you expect them to give you? What additional information will you need by the time you reach the hospital?

____

____

____

____

____

CONTENT APPLICATION #2

Your hazmat response team has been called to the hospital where a gallon of formaldehyde has been spilled in the laboratory. You are familiar with the facility because you assisted in the training of their in house hazmat brigade. When you arrive, a maintenance supervisor, who is the brigade leader, advises you that the area has been closed off and evacuation of the area involved is completed. What additional questions should you ask of the supervisor? What other jobs would you expect to be accomplished prior to your arrival?

____

____

____

____

____

CHAPTER 8 SCENARIO

You have been dispatched to the trauma center in your community at the request of the hazardous materials unit on the scene of an overturned tanker truck. Because you work in the hospital emergency room on your off days and are familiar with their routine and are also familiar with hazmat response, the team wants you to coordinate efforts with hospital staff in the preparation of receiving hazmat patients. The truck is located on an interstate highway mostly used by weekenders returning from a day at the beach. The information received from dispatch indicates that the tanker was carrying 4,500 gallons of a high concentrate parathion to a local flying service where it was to be diluted and spread over crops, rolled over and ruptured, spilling the contents over the road. Many vehicles traveling home from the beach were held up in the spill for approximately 20 minutes before being directed by the local police to drive on through to reach their destination. The concern of the hazmat team is that many people received a significant exposure and will probably report to the hospital once symptoms surface. Furthermore, the team was also treating two officers that became very ill after apparently standing in the spill while they directed traffic. It is Saturday, June 5th, the temperature is 85° F with a humidity of 78%.

When you arrive at the hospital, you notice a group of about fifteen individuals being held outside of the emergency room entrance by two nurses dressed in gowns and particulate masks. All of the 15 people are ambulatory but displaying signs of organophosphate poisoning (excessive lacrimation and salivation). Most are dressed in swimming suits and tee shirts.

Scenario Questions

1. Is a decontamination system necessary for these patients?

___

2. If so where would the best place be to set one up? What equipment is needed?

___

___

___

3. Is the capture of runoff important for this emergency decontamination?

___

4. Will accessing the hospital's hazardous materials contingency plan give us all of the information needed to adequately decontaminate these people?

___

___

___

___

5. Knowing the possible routes of entry for the chemical may be important. What are some of the means available in the hospital to get this information?

___

___

___

___

6. Is a command structure needed to handle this hazmat event at a hospital?

___

___

___

___

7. What are some of the other concerns for the emergency room when contaminated patients arrive there?

8. Is the protective equipment used by the initial two nurses adequate? What should be added or deleted?

9. Utilizing the in-house resources, where can more advanced personal protective equipment come from?

10. Once these patients are through a washing process are they free of contamination? Explain?

11. Is your hospital prepared to handle an incident of this magnitude?

12. Does your hospital have written SOPs and a plan to activate if contaminated patients arrive?

13. Does your hospital train with the hazardous materials response team so that an operation of this type is controlled and mitigated without total confusion or chaos?

Biohazard Awareness, Prevention, and Protection

Objectives

Given an incident involving biohazardous materials, the student should be able to identify the unique hazards involved, including the means of entry, virulence, infectiousness, invasiveness and pathogenicity. The student should also be informed about the regulations involving biohazardous waste, the Ryan White Act, and exposure reporting.

As a hazardous materials medical technician, you should be able to:

- Identify the difference between bacteria and a virus.
- Define biohazard.
- Define infection control.
- Identify the means of entry to include the:

  Virulence — Invasiveness

  Infectiousness — Pathogenicity
- Identify the key points to the Ryan White Act.
- Identify the steps to clean up of a biohazard site.
- Formulate a high level, intermediate level, and a low level disinfectant and define the use of each level.

CONTENT QUESTIONS

1. Explain the difference between bacteria and viruses.

Answer questions 2–7 using the following terms:

A. invasiveness
B. infectiousness
C. primary pathogens
D. infection
E. pathogenicity
F. opportunistic pathogens

2. ____________________ is the result of either bacteria or virus gaining access into the body and multiplying. This invasion overwhelms or bypasses the defensive barriers always present in a healthy host.

3. ____________________ can, unaided by other opportunities, invade a healthy body, bypassing defensive mechanisms and establishing an infection.

4. ____________________ are usually unable to penetrate the defensive mechanisms found in a healthy individual.

5. ____________________ is the ability, once a pathogen gains access into the body, to initiate and maintain an infection.

6. ____________________ is the ability of the pathogen to progress further into the body once an infection is established.

7. ____________________ is the ability of the pathogen to injure the body once infection is established.

8. When using the terminology "universal precautions" what is meant by it and how does it apply to patient care?

____

____

9. The Ryan White Act allows for communication with the hospital when an exposure takes place through
   A. the E.R. physician.
   B. the infection control nurse.
   C. the designated officer.
   D. the exposed emergency caregiver.

10. HEPA stands for
    A. highly effective particle absorption.
    B. hard extraction of particle air.
    C. heavy extraction and particle absorption.
    D. high efficiency particulate air.

11. HEPA masks or respirators certified under 42 CFR part 84 subpart K are required for all the following except
    A. when workers enter rooms housing individuals with suspected or confirmed infectious TB.
    B. when workers are present during the performance of high-hazard procedures on individuals who have suspected or confirmed infectious TB.

C. when emergency medical response personnel or others transport, in a closed vehicle, an individual with suspected or confirmed infectious TB.

D. before evaluating individuals who are complaining of a productive cough over several weeks.

12. A significant exposure is defined as:

____

____

13. The use of physical or chemical means to remove, inactivate, or destroy bloodborne pathogens on a surface or item to the point that they are no longer capable of transmitting infectious particles, and the surface or item is rendered safe for handling, use, or disposal is the definition of

A. routine equipment cleaning.
B. biohazard decontamination.
C. medical tool disinfection.

14. There are four levels of decontamination indicated for killing microbial agents. Which of the following is not a level of decontamination?

A. Sterilization
B. Low level disinfection
C. High level disinfection
D. Surface disinfection
E. Intermediate level disinfection

15. What chemical is known to elongate latex gloves, enhancing penetration of liquids through the glove pores?

A. Solvents
B. Blood
C. Isopropyl alcohol
D. Bleach

16. Placing blood soaked clothing into evidence bags where the clothing may be left for extended periods of time may cause what problem? What is the danger of this practice?

____

____

____

CONTENT APPLICATION #1

You have just reported for duty on Rescue 11 and have been told that a small aircraft crashed in your territory during the night. There were no survivors due to the speed of the airplane when it made contact with the ground. An investigation of the crash continued throughout the night and you will be going out this morning to provide whatever extrication will be needed to remove the bodies. Knowing that the scene will be grossly contaminated with blood and body fluids, what personal protective equipment will be needed to protect yourself and the crew? How will the equipment be decontaminated after the event? Should protective respiratory masks be used during the extrication and removal of the bodies? Explain.

___

___

___

CONTENT APPLICATION #2

You have just returned to the station from a call involving an assault. The patient to whom your crew was rendering care had been struck in the face with a bottle. He was obviously intoxicated and did not care whether the crew assisted him or not. When attempting to interview him, the patient spit a mixture of blood and sputum into one of the crew's face. Droplets entered the crew member's mouth and eyes. Is this considered a significant injury? What steps must be taken to determine if the patient is infected with a bloodborne disease? Can this patient be tested at your department's expense? Explain.

___

___

___

___

___

CHAPTER 9 SCENARIO

It is 3 A.M. and you have just received a call to the local homeless shelter where there has been a recent outbreak of tuberculosis. The call this evening is for another patient complaining of shortness of breath. Just last week you transported a patient who complained of fever, night sweats, and shortness of breath. Later you were advised that the patient was infected with TB and the infection was active. Being concerned for your own health you reported the exposure and were tested with a PPD (purified protein derivative). So far the tests have been negative and the thought has passed until now. In the homeless shelter, several hundred residents sleep in close proximity to one another. Many are in poor health due to a lack of medical care. Your experience dealing with this group tells you that many communicable diseases have been found here from AIDS to hepatitis. Unfortunately, the symptoms of TB seem to activate at night stimulating many emergency responses to the facility.

When you reach the patient, he is profusely sweating and coughing. He appears in poor general health. Gathering a history reveals that the patient has been living outside for the past 6 months, is a heavy drinker and smoker. A physical exam reveals warm, wet skin with poor turgor. Vital signs are: BP 140/88, pulse 108, respirations at 24, and a SaO_2 of 88%. Lung sounds indicate rales in both bases with rhonci in the upper airways, slightly diminished in the right lower lobe.

Scenario Questions

1. What precautions should be taken with this patient?

____

____

____

2. At what time during this call should these precautions be taken?

____

____

____

3. Is there a concern with transporting this patient?

____

____

____

4. After transporting this patient should the equipment be cleaned? How?

____

____

____

5. What solution should be used to clean items such as the stretcher? BP cuff? Walls of the unit?

____

____

____

6. How does tuberculosis enter the body?

____

____

____

7. Before you were able to protect yourself, the patient coughed in your direction causing sputum to strike your face and clothing. How would you clean your skin? Clothing?

___

___

___

8. Is this exposure significant?

___

9. Do you have the right to be informed if the patient later tests positive for a communicable disease?

___

10. Do you have the right to be tested for the disease if the patient tests positive?

___

11. Does your department have an infection control policy? If so are you familiar with the documentation needed after a significant exposure?

___

___

12. Can you return to the hospital tomorrow to find out what was wrong with the patient? If not, how do you find out?

___

___

___

Clandestine Drug Laboratories

Objectives

Given a hazardous materials incident involving a clandestine drug laboratory, the medical responder should be familiar with the processes involved in making illicit drugs. The responder should also be familiar with the chemicals involved and the types of injuries that may be incurred by the operator of the lab, police, fire, or other official responder who ventures in without knowledge of the dangers involved with such laboratories.

As a hazardous materials medical technician, you should be able to:

- Understand the scope of the problem involving the manufacturing of illicit drugs throughout the United States.
- State what are the most common types of booby traps found in conjunction with these laboratories.
- Identify possible health and safety hazards found at clandestine drug laboratories to include but not limited to:

  Chemical hazards: acids and bases
  Metal poisons
  Flammables
  Poisons
  Respiratory Irritants
- Identify the most common types of labs found and the dangers presented by each.

  Amphetamines and methamphetamines
  Phenocyclidine (PCP)

CONTENT QUESTIONS

1. Explain why methamphetamine drug labs are expected to increase during the next several years.

____

____

____

2. Protection of clandestine laboratories is done through several means. Name 4 of the most commonly found booby traps.
   1. ____
   2. ____
   3. ____
   4. ____

3. Many of the drug labs are discovered as a result of police investigations. According to the author, it is estimated that ______ % are discovered as a result of explosion or fire.

   A. 10 **B.** 20 **C.** 30 **D.** 40

4. Name 6 locations where drug labs have been discovered.
   1. ____
   2. ____
   3. ____
   4. ____
   5. ____
   6. ____

5. Explain the workings of a foil bomb.

____

____

____

6. Hydrochloric acid and NaCN forms what when mixed?

   A. Hydrogen fluoride
   B. Sodium nitrile
   C. Hydrogen cyanide
   D. Hydrogen nitrochloride

7. The author lists three things that a clandestine drug lab is. Name them.
   1. ____
   2. ____
   3. ____

8. Emergency medical responders often know minute details about their response territories. This in-depth knowledge may clue them into recognizing an operating drug lab. What clues might the medical responder notice that would indicate a possible lab? Name 6 clues.
   1. ______
   2. ______
   3. ______
   4. ______
   5. ______
   6. ______

9. All of the following may be found at a drug lab except:
   A. Respiratory irritants
   B. Corrosives
   C. Flammables
   D. Radioactive material
   E. Poisons

10. An arrested lab operator states that he is only making a "few jelly beans and double crosses." What do you expect he is talking about?

______

______

______

______

______

11. Once a methamphetamine lab is discovered by an emergency medical responder, the first action by that individual should be to ______. (Explain all the steps completely.) *RUN is not an acceptable answer.*

______

______

______

______

______

CONTENT APPLICATION #1

You are dispatched to a routine medical call (the patient is complaining of shortness of breath). Once you arrive, your partner begins to assess the patient. Vital signs are taken and information gathered to treat the patient. Because the patient is in distress, she is unable to state what medications she is taking. You start looking through the house for the patient's medication and you discover several pieces of unusual glassware being heated in the bathroom, and a yellow liquid in one of the flasks. You believe this to be an illegal drug laboratory. What steps should be taken next? Is it safe to remain in the house? Why?

____

____

____

____

____

CONTENT APPLICATION #2

You have been dispatched to an auto accident. When you arrive on the scene, a police officer informs you that the auto was carrying chemicals for use in making illegal drugs. Several of the containers were broken in the crash including one that smells of cat urine. Others are spilled inside the vehicle. What kinds of chemicals would you expect to be transported for use in a clandestine laboratory? What dangers exist to those workers exposed to the chemicals inside the auto? What agencies should be contacted about chemicals of this nature?

____

____

____

____

____

CHAPTER 10 SCENARIO

You are manning a fire department-based ALS nontransport rescue unit. You have just arrived on the scene of a shooting involving a 22-year-old male. Police, who were on the scene when you arrived advise you that the patient is in the living room. When you reach the patient, a strong odor of solvent is noted on his clothing. The gunshot wound is to the lower leg and is not life threatening. While you are splinting and dressing the wound and making the patient ready for transport, one of the police officers asks you to assess something she found in the back bedroom. When the ambulance personnel arrive you transfer the patient care to them and meet with the police officer.

The police officer directs you to the bathroom off of the master bedroom. The bathroom contains several large glass bottles, heating elements, condenser tubes, and collection containers. A strong odor of solvent and an irritating acid type taste is noted when you breathe. Having trained with the hazmat team you recognize that this is probably a drug laboratory. Furthermore, you advise the police that the laboratory appears to be functioning. Knowing the inherent dangers of a clandestine laboratory you advise the police to evacuate the building and call for the hazmat response team.

Scenario Questions

1. In addition to the obvious chemical hazards, what other kinds of hazards may be present?

____

____

____

2. Is it important to question the injured patient about the process?

____

____

3. Because you entered the cooking room and were able to detect the process by smelling the air, is this considered an exposure?

____

____

4. Is it true that small labs of this type usually do not possess a hazard?

____

____

5. If the patient advises that methamphetamine is the drug being prepared, what are some of the chemical hazards?

____

____

____

6. Is there a danger of a fire at a laboratory such as this?

____

____

7. If you are asked by the police to turn off the chemical process, what is your reply? Is this a decision you are prepared to make? Explain.

____

____

____

____

8. Should the bomb squad respond to the scene? Why?

9. If this emergency response happened to you today, whose responsibility is it to transport the chemicals away from the scene and who will dispose of them?

10. If this took place in your district, once the chemicals are confiscated and transported, whose responsibility is it to clean the house to ensure safety?

Air Monitoring

Objectives

When presented with an incident, the student should be able to define, employ consideration factors, and demonstrate initial monitoring procedures as they pertain to scene stabilization.

As a hazardous materials medical technician, you should be able to:

- Describe the four hazardous environments that air monitoring is directed towards.
- Discuss how meteorological conditions affect the monitoring process.
- Describe the difference in general air monitoring techniques and the considerations thereof.
- Describe mathematically the association between parts per million and percent.
- Discuss the physical and chemical properties that affect air monitoring.
- Discuss oxygen deficiency monitors.
  Limits of oxygen deficiency and enriched.
  Oxygen gradients.
  Procedures when utilizing the oxygen deficiency monitor.
  How the oxygen deficiency monitors work.
  Limitations of oxygen deficiency monitors.
- Discuss colormetric tubes.
  Mechanics of operation.
  Procedures when utilizing colormetric tubes.
  Three measurement methods.
  Limitations of colormetric tubes.
- Discuss photoionization detectors.
  Mechanics of operation.
  Relative response patterns.
  Limitations of the photoionization detector.

- Discuss flame ionization detectors.
  Mechanics of operation.
  Limitations of the flame ionization detector.
- Discuss combustible gas indicators.
  Mechanics of operation.
  Response curves or reference charts.
  Limitations of the combustible gas indicator.
- Discuss radiation detectors.
  Mechanics of operation.
  Limitations of the radiation detector.
- Discuss the level for personal protective equipment for use with air monitoring.

CONTENT QUESTIONS

1. Air monitoring is a technique that will enable the rescue worker
   - **A.** to establish the known environmental conditions.
   - **B.** to give the medical sector information on the environment.
   - **C.** to have the ability to gather information.
   - **D.** to do all the above.

2. This collection of data will assist the emergency worker in the decision making process
   - **A.** when it comes to entry team treatment.
   - **B.** when it comes to the treatment of potential victims of the incident.
   - **C.** utilizing both A and B.
   - **D.** utilizing neither A nor B.

3. Air monitoring is a technique that is not without is own limitations and pitfalls.
   - **A.** True
   - **B.** False

4. 29 CFR 1910.120 (h) states specific requirements for air monitoring procedures.
   - **A.** True
   - **B.** False

5. There are four basic considerations when dealing with air monitoring devices:
   - **A.** Oxygen deficiencies, explosive atmospheres, flammability, radoactivity
   - **B.** Oxygen concentrations, toxic atmospheres, explosive conditions, radioactive ionization
   - **C.** Oxygen concentrations, flammability, radioactive ionization, biohazardous atmospheres
   - **D.** Oxygen concentrations, toxic atmospheres, explosive conditions, biohazardous atmospheres

6. Oxygen-deficient atmospheres can be classified as:
   - **A.** Simple, as the displacement of oxygen by other gases
   - **B.** Immediate hazard, which depletes the oxygen concentration, leading to asphyxiation
   - **C.** Inappropriate ventilation techniques, or unventilated areas
   - **D.** A and B

7. Toxic atmospheres are
   - **A.** atmospheres in which the chemical is a vapor, mist, fog, dust, fume and/or aerosol.
   - **B.** environments that are immediately dangerous to life and/or health.
   - **C.** usually found under or around the LEL.
   - **D.** all of the above.

8. In air monitoring of explosive conditions
   A. be sure the air monitoring instrument in use is rated for entry into these types of atmospheres.
   B. be sure the instrument in use has the Factory Mutual Research Corp. (FM), Underwriters Laboratory (UL), or a recognized European certification.
   C. both A and B are incorrect.
   D. both A and B are correct.

9. Which statement about radioactive isotopes is false?
   A. Radioactive isotopes are found within any community.
   B. Alpha and beta particles are respiratory hazards.
   C. Gamma waves can cause damage to tissues and organs.
   D. Radioactive materials usually are affected by meteorological conditions.

10. Weather conditions can affect our air monitoring procedures.
    A. True B. False

11. Wind speed and direction can affect your scene in two distinct ways.
    A. By spreading contaminates downwind.
    B. In combination, they can greatly affect the chemical and physical properties of the chemical(s) we are evaluating.
    C. The possibility of wind direction change and setting up the initial command and support functions must be considered.
    D. A and C

12. All instruments are rated and function at optimal temperatures.
    A. True B. False

13. Temperature fluctuations
    A. can change a chemical's properties.
    B. can increase vapor pressure, increasing the probability of reaching our dangerous LEL.
    C. can saturate the mechanism and give false readings.
    D. can decrease the possibility of permeation.

14. Humidity can affect chemicals. Which is a false statement?
    A. Humidity can greatly influence the frequency of heat stroke.
    B. Air monitoring will become increasingly difficult.
    C. Air moisture can mix with vapors of a corrosive chemical causing injuries.
    D. The filters and burning chambers of an air monitoring device work at a greater efficiency under high humidity conditions.

15. An increase in humidity can cause moisture within the monitoring device. This saturation will distort the instrument's ability to accurately read the environment being monitored.
    A. True B. False

16. Temperature dew point
    - **A.** is that temperature at which the moisture in the air will become visible.
    - **B.** is the water saturation within the air at a certain atmospheric pressure.
    - **C.** and the ambient temperature have a gradient.
    - **D.** statements are all correct.

17. The larger the numbers (the difference between dew point and the temperature), the less saturated the air is with water.
    - **A.** True
    - **B.** False

18. The same problems we had with temperature and humidity will become evident in situations that create a decreasing temperature toward the dew point.
    - **A.** True
    - **B.** False

19. Fog will start to appear within _____° of dew point:
    - **A.** 3–6
    - **B.** 4–8
    - **C.** 4–5
    - **D.** 1–4

20. With barometric pressure,
    - **A.** as pressures increase, the potential for evaporation will decrease.
    - **B.** as the atmospheric pressure decreases, the evaporation of chemicals will increase.
    - **C.** A only or B only are true.
    - **D.** both A and B are true.

21. Which statement is false about air monitoring devices?
    - **A.** Interference that registers a high reading is called positive interference.
    - **B.** Low readings are a result of negative interference.
    - **C.** Air monitors can distinguish between chemicals.
    - **D.** The technician must be able to understand the information the air monitor provides.

22. The definition of interference gases is:
    - **A.** Gases that may be present that are similar to the gases being tested.
    - **B.** Registering a high reading is called the positive interference and is similar to calibrant gases.
    - **C.** Low readings are a result of negative interference and are similar to calibrant gases.
    - **D.** All of the above.

23. Give an example of a common interference gas:

    ____

    ____

    ____

    ____

24. There are basically two types of monitoring techniques. Which of the following best describes the type of air monitoring that one would employ at the scene of a hazardous materials incident or a confined space operation?
   - A. Direct reading instruments
   - B. Real time monitoring
   - C. Sample analysis
   - D. Sample collection

25. Describe each of the above four responses.

____
____
____
____
____
____
____
____

26. DRI is an abbreviation for
   - A. Direct reading instrument.
   - B. Deliberate reading instrument.
   - C. Direct response instrument.
   - D. Direct release instrument.

27. DRIs will identify any chemical in question.
   - A. True
   - B. False

28. False readings are abnormal for DRIs
   - A. True
   - B. False

29. Numbers are useless unless we know how to apply them and what they mean.
   - A. True
   - B. False

30. Describe concentration.

____
____
____
____
____

31. ppm is an abbreviation for: ____

32. Give a visual representation of what ppm is describing.

____
____
____

33. How can the shape, velocity, or size of a molecule affect its toxicity?

____

____

____

34. 7.7% describes how many parts per million?

A. 77 ppm **B.** 770 ppm **C.** 7,700 ppm **D.** 77,000 ppm

35. What is the definition of "boiling point"?

____

____

____

36. What is the definition of "state of matter"?

____

____

____

37. What is the definition of "melting point"?

____

____

____

38. What is the definition of "specfic gravity"?

____

____

____

39. What is the definition of "vapor density"?

____

____

____

40. What is the definition of "vapor pressure"?

____

____

____

41. What is the formula we can use to algebraically give ppm from mg/m^3?

42. What is the definition of "solubility"?

43. What is the definition of "flammable range"?

44. Describe the four basic environmental atmospheres that we are going to consider.

45. Describe how pressure can affect oxygen concentrations.

46. Describe the process (technique) that should be employed when testing an atmosphere.

47. What are the OSHA limits for an oxygen-deficient and an oxygen-enriched atmosphere?

A. 20.0% and 21%

B. 19.5% and 23.5%

C. 19.0% and 21%

D. 19.5% and 22.5%

48. Describe why OSHA's definition of an oxygen-deficient or an oxygen-enriched atmosphere is potentially dangerous.

49. What should you do if an audible alarm sounds on an oxygen monitor?

50. How does an oxygen sensor work?

51. What chemicals should be considered as contaminants to an oxygen sensor and why?

52. What are the limitations of an oxygen sensing instrument?

53. Name three occurrences that may be interpreted by the instrument.

54. What three instruments can be used to detect a toxic atmosphere?

55. Colormetric tubes will identify a chemical or chemical class. Explain.

56. Describe how a colormetric tube works.

57. Describe the different measurement scales that may be found on a colormetric tube.

58. What technique must be employed when utilizing a colormetric tube for atmospheric monitoring?

59. In general, detector tubes indicate measurement in three ways. Name these three ways, and the limitations of each.

60. Describe how temperature and humidity affect the chemical reactions that take place in colormetric tubes.

61. Describe the limitations inherent in colormeteric tubes.

62. How does the PID work, and what does PID mean?

63. Describe the elements that can interfere with the operation of the PID.

64. List the chemicals that are used to create ionizing ultraviolet light, and their respective eVs.

65. How do we read the PID?

66. What are the limitations of the PID?

67. How does the FID work, and what does FID mean?

68. Describe the interference that can occur with the FID.

69. How do we read the FID?

70. What are the limitations of the FID?

71. How does the CGI work, and what does CGI mean?

72. Describe the problems that can occur with the CGI.

____

____

____

____

____

73. How do we read the CGI?

____

____

____

____

74. What are the limitations of the CGI?

____

____

____

____

____

75. Radiation is classified into two types of energy. What are they?

A. Radioactive
B. Ionizing radiation
C. Non-ionizing radiation
D. A and B
E. B and C

76. What type of energy is radiation?

____

77. Describe alpha particles.

____

____

____

78. Describe beta particles.

____

____

____

79. Describe gamma radiation.

---

80. Describe how a Geiger counter measures radiation.

---

81. Describe how a dosimeter works.

---

82. Describe the units of measurement for radiation.

---

83. What three radiological measurements can be considered equal in an emergecy situation?

---

84. Describe the limitations of radiological monitors.

85. What PPE practices should one employ when performing air monitoring?

____

____

____

____

____

86. Describe the difference between testing and calibration.

____

____

____

____

____

____

CONTENT APPLICATION

You are on the scene of a hazardous materials incident in which chemicals have been released within a structure. Within this occupancy, which is three stories high with a common HVAC system, the following chemicals have been released:

Isopropyl alcohol
Hexane
Sulfur dioxide
Methyl ethyl ketone

A reconnaissance team has come back with the following information: colormetric tubes do not indicate the presence of sulfur dioxide, PID readings do not identify or deny any possible toxic environment. The team reports 100% LEL and 22% LEL fluctuations, 40 ppm of CO, and 20.0% oxygen. Describe what is possibly occurring with your instrumentation.

CHAPTER 11 SCENARIO

You are on a scene of a hazardous materials incident. You have been placed in charge of the air monitoring sector. So far the information that has been given to you has been very sketchy. The wind is from the southwest at 2 mph, with a temperature of 88°, and a humidity factor of 50%. Answer the questions below as to how you will appropriately monitor for contaminates.

Scenario Questions

1. How will the given weather conditions limit your air monitoring process, or will they?

2. If this incident occurred in the winter with temperatures of 40° and wind speed of 5 mph, how would monitoring change?

3. If this incident occurred in the summer with temperatures of 95° and humidity of 80%, without any wind, could this affect the process and how?

4. Given the above scenario with chlorine as an interference gas, what would be your plan of action?

5. Describe the limitations of the following:
   A. Oxygen deficient monitors

   B. Colormetric tubes

   C. Photoionization detectors

D. Flame ionization detectors

E. Combustible gas indicators

F. Radiation detectors

6. What level of personnel protection should one be in when providing air monitoring?

7. Discuss the considerations of air monitoring in terms of patient contact.

Confined Space Medical Operations

Objectives

Given a confined space incident, the student should be able to identify the appropriate level of entrance along with rescue and safety procedures.

As a hazardous materials medical technician, you should be able to:

- Identify the laws surrounding permitted and nonpermitted entrance into the confined space.
- Define confined space and OSHA's four criteria for permitted entrance.
- Discuss scene management as it pertains to confined space entry.
- Discuss the atmospheric hazards that may be present, identifying each concern and associated hazards.
- Discuss the physical hazards that may be present, identifying each concern and associated hazards.
- Describe the similarities between the hazardous materials accident and the confined space response.

CONTENT QUESTIONS

1. The Federal Code that identifies the requirements for confined space operations is:

   A. 29 CFR 1910.120
   B. 29 CFR 1910.134
   C. 29 CFR 1910.100
   D. 29 CFR 1910.146

2. Identify the false statement.

   A. In all advanced rescue operations, preplanning and extensive training is the avenue for safe operations.
   B. The confined space rescue team must simulate a rescue operation within one of the preplanned areas annually.
   C. The confined space rescue team must train in an area that would be representative of the preplanned spaces.
   D. Each member must have a minimum of CPR and standard first aid.

3. Entry into a confined space can be defined as, once the rescuer breaks the plane into the space, entry has occurred.

   A. True
   B. False

4. The rescuers shall have the appropriate level of personnel protection; however, they do not have to provide air monitoring capabilities.

   A. True
   B. False

5. You are called to a confined space incident and have identified that a safe rescue can be performed given an appropriate level of protection. All confined space regulations have been employed (29 CFR 1910.146). However, this incident is at a city construction site. Which are true statements?

   A. If you follow the standard 29 CFR 1910.146, all legal obligations have been addressed.
   B. Following an additional standard may be necessary.
   C. Utilizing 29 CFR 1910.120 and 29 CFR 1910.146, all legal obligations have been addressed.
   D. B and C

6. The establishment of working SOGs, associated with preplanning and training will have to be done prior to confined space rescue operations.

   A. True
   B. False

7. OSHA has established criteria for the confined space. This space must meet all of these criteria for it to be called a "confined space." Describe these criteria:

   ____

   ____

   ____

   ____

   ____

8. According to the OSHA standard there are four basic criteria that establish the need for a permit. Which of the following is a false statement?
   - **A.** The space has an atmosphere that is hazardous or has the potential to become hazardous.
   - **B.** The internal configuration of the confined space is such that a person within the space is safe, once entered.
   - **C.** The material within the confined space can or has the potential possibility of victim engulfment.
   - **D.** The presence of any other recognizable hazard which may cause serious safety and/or health effects.

9. For the above question, describe the false statement and discuss the appropriate verbiage:

   ____

   ____

   ____

   ____

   ____

10. Who is responsible for the identification of a permitted or nonpermitted confined space?
    - **A.** Fire department official
    - **B.** Facility representative
    - **C.** OSHA representative
    - **D.** Both A and B

11. Once a space has been identified as a permitted confined space, what must occur?
    - **A.** Hazards that are directly affecting the rescue attempts must be controlled.
    - **B.** All hazards must be identified and controlled before entrance can take place.
    - **C.** Air monitoring only should precede any entry.
    - **D.** Personal protective equipment must control all referenced hazards.

12. Once a confined space is reclassified as a nonpermitted space, *all* OSHA standards apply.
    - **A.** True
    - **B.** False

13. All confined spaces should be downgraded to nonpermitted spaces as soon as possible.
    - **A.** True
    - **B.** False

14. At the permitted level of confined space, a high degree of personnel and equipment are required. This would be analogous to a Level A entry within a hazardous materials incident.
    - **A.** True
    - **B.** False

15. The only way a reclassification can occur is if all hazards have been removed for the entire duration of the incident.
    - **A.** True
    - **B.** False

16. The documentation of a confined space downgrade is not required.
   A. True B. False

17. A cursory medical is not necessary at the confined space operation.
   A. True B. False

18. Lockout and tag-out are terms that describe which statements?
   A. A procedure used to eliminate electrical and mechanical hazards within a confined space only.
   B. A procedure by which the victim is managed.
   C. A procedure by which the hazards are isolated and personnel are identified.
   D. A procedure by which isolation of the hazard occurs with identification of the victim.

19. There is no reason to identify the areas of a confined space operation like we do at the hazardous materials event, i.e. hot, warm, and cold zones.
   A. True B. False

20. Describe the hazards that may be found within the confined space.

   ____

   ____

   ____

   ____

   ____

21. Within a confined space operation, we can identify the areas of scene similarily to a hazardous materials operation. Identify the following:
   A. Warm zone ________ 1. The confined space itself
   B. Cold zone ________ 2. Support team's placement
   C. Hot zone ________ 3. Buffer zone around the immediate hazard

22. Confined space operations do not need information about the rescue, hazards involved, or the management of the patient.
   A. True B. False

23. Communication at the confined space operation can occur with hand signals only.
   A. True B. False

24. Communication should be
   A. a combination of visual contact, hardline (radio) verbal communication, and hand signals.
   B. via radio alone which will provide a safe operation.
   C. taken over by the radio once the rescuer is beyond the visualization of the safety officer.
   D. unnecessary during the rescue. All it does is confuse and distract the rescuer.

25. Discuss each of the above responses in question 24.

____

____

____

____

____

____

____

____

26. All personnel that enter the hot and warm zones must receive a cursory medical examination.

A. True **B.** False

27. Describe the reasoning behind your answer for question 26.

____

____

____

____

____

28. Preplanning of all known "spaces" and potential ones should have standard operating procedures drawn up.

A. True **B.** False

29. The management qualities of a commander should include

A. having strong character and knowledge of the tasks to be accomplished.

B. ensuring accountability and safety.

C. being the facilitator of the incident.

D. all of the above.

30. The goals of the commander are

A. to remove the victim(s).

B. to accomplish this in a timely manner.

C. to conserve life and property.

D. all of the above.

31. Evaluate, revise, and identify is a continuous process by the commander at all incidents. Anticipating future needs and addressing these issues constantly becomes a part of this process.

A. True **B.** False

32. Describe how the visual safety officer and the incident commander will interact.

33. Describe the prime responsibility of the visual safety officer.

34. Describe the other supportive functions that can occur at the confined space operation.

35. Describe the potential hazards that can occur within the immediate area around the confined space.

36. Air monitoring is only necessary if there is an identifiable hazardous chemical involved.

A. True B. False

37. Discuss why air monitoring is important.

38. Describe the hazard categories and their subparts.

39. What is the average oxygen concentration within ambient air?

A. 21% **B.** 20.95% **C.** 20.7% **D.** 21.5%

40. OSHA has established that an oxygen deficient atmosphere is below a certain percent, and in an enriched atmosphere oxygen concentrations are above a certain percent. Which percentages are correct?

A. 19.5% and 23%
B. 19% and 23%
C. 19.5% and 23.5%
D. 19.5% and 23%

41. Describe how oxygen pressures could affect a patient.

42. Describe problems associated with enriched oxygen atmospheres.

43. Describe medication that could adversely affect an individual in an oxygen enriched atmosphere.

44. Describe the causes of oxygen deficiency within a confined space.

45. How would an enriched oxygen atmosphere affect air monitoring?

46. How would temperature affect the air monitoring process?

____

____

____

47. Explain the following statement.

Oxygen content lower than 20.7 should be considered as oxygen deficient or at an "IDLH" level.

____

____

____

____

____

48. Discuss the primary difference between flammable and combustible.

____

____

____

49. Discuss the relationship between temperature, flash point, and vapor pressure.

____

____

____

____

50. How would the above information (questions 48 and 49) relate in the real world?

____

____

____

____

51. 1% of a 50,000 gallon tanker is:

A. 5000 **B.** 500 **C.** 50 **D.** 5

52. If acidic and alkali solutions are allowed to come into contact with each other, a reaction takes place; or if used in certain containers, will produce

A. flammable gas.

B. toxic gas.

C. both A and B.

D. none of the above.

53. The range of flammability is described by two points. What are these points?

A. LEL and UEL

B. IDLH and LEL

C. IDLH and UEL

D. WFP and UEL

54. What do the abbreviations you have chosen in question 53 stand for?

A. Lower explosive level

B. Upper explosive level

C. A and B

D. None of the above

55. Describe the chemical groups that have wide flammable ranges.

____

____

____

____

____

56. What is the difference between wide flammable ranges and small flammable ranges, and why is this information important?

____

____

____

____

____

57. Flammable ranges are stable and absolute values that can be referenced and established.

A. True

B. False

58. What are your other considerations in terms of flammability within a confined space?

____

____

____

____

59. Give the technical definition of an explosion.

____

____

____

____

60. Describe the explosion in terms of two distinct occurrences and the four categories of injuries.

____

____

____

____

61. Describe what would occur if someone was subjected to a blast wave.

____

____

____

____

62. To have a margin of safety, in all confined space environments that exhibit an LEL greater than __________ , entry is prohibited.

A. 1% B. 5% C. 10% D. 15%

63. Where are the toxic levels in relation to the flammable range?

A. Within the flammable range
B. Below the flammable range
C. Above the flammable range
D. Both A and B are correct

64. Describe how electrical hazards can be controlled.

____

____

____

____

65. Describe how mechanical hazards can be controlled.

____

____

____

____

66. Describe how structural hazards can be controlled.

____

____

____

____

67. Describe in terms of function how a confined space operation should be set up.

______

______

______

______

______

68. Describe how ventilation can help in a confined space operation, and how it can hinder the operation.

______

______

______

______

______

69. Describe the benefits of a cursory medical at the scene of a confined space operation.

______

______

______

______

______

70. Describe the function of the tactical officer.

______

______

______

______

71. Describe the additional dangers that may be presented to the rescue team in terms of patient care, and the application of treatment within the confined space.

______

______

______

______

72. Discuss briefly decontamination at the confined space operation.

____

____

____

____

73. Discuss the rehabilitation sector at the confined space operation.

____

____

____

____

CONTENT APPLICATION

You have been called to a sewer transfer facility. Within this facility, sewage is pumped from the surrounding neighborhoods and moved on to the county's treatment plant. You have knowledge of the structure and the possible hazards. They have had, on several occasions, a build-up of hydrogen sulfide gas. Pumps are always active, with a standby generator. The following is a schematic of the facility:

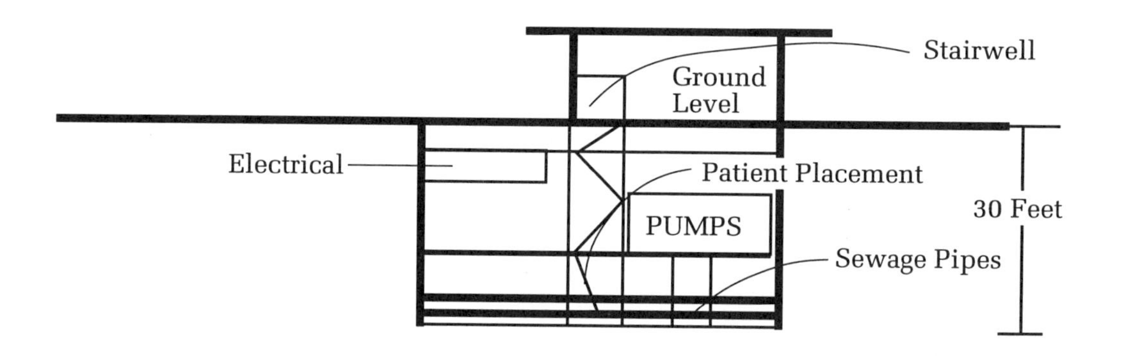

You have been advised by dispatch that the patient is at the bottom of a steep stairwell. How would you manage this incident as a rescue, and what would be your considerations in terms of incident management, air monitoring, patient removal, supplied air, ventilation, lighting, lockout and tag-out?

CHAPTER 12 SCENARIO

You and your partner are on the hazardous materials team as hazmat medical technicians. Your unit has been requested to respond to a confined space operation. You recognize the address and start to discuss the facility with your partner. She remembers that there is a 50,000 gallon container that was emptied and placed underground many years ago. Both of you try to remember the original contents of the container, but cannot.

Your prehazard plans do not describe this vessel and actually state that it has been removed per the property owner. However, the dispatcher updates the incident with a description of this vessel as the confined space. All you and your partner can remember is that it was once a container that had flammable liquids. Your discussion progresses to the weather and how hot it has been lately. The temperature is 90° F with a humidity factor of 60%.

Scenario Questions

1. Considering that this vessel may have contained a volatile chemical at one time, what are your considerations?

____

____

____

____

2. How will you set up a sectorization of such an incident?

____

____

____

____

3. What are your concerns within each sector?

____

____

____

____

4. How can the incident management system assist your operation?

____

____

____

____

- Given a small incident can you do away with the incident management system?

____

- Under what conditions can you classify the confined space as a nonpermitted space?

____

____

____

____

5. List the hazard atmospheres.

____

____

____

____

____

6. What consideration does each hazard atmosphere present to the medical technician?

____

____

____

____

____

7. What are your considerations at such an incident?

____

____

____

____

____

Acronyms

Much of the jargon in the Emergency Medical Services and the Hazardous Materials field are words which are formed by the initial letters of each word in the phrase or clinical terminology. The following is an alphabetical listing of such acronyms and jargon used in emergency response:

ABC	Airway, Breathing, Circulation
A/C	Air-conditioning
ACGIH	American Conference of Governmental Industrial Hygienists
ADA	Americans with Disabilities Act
ALD	Average Lethal Dose
ALS	Advanced Life Support
APIE	Analyze, Plan, Implement, Evaluate
APR	Air Purifying Respirator
ATP	Adenosine triphosphate
AV	Arterioventricular
BEK	Butyl ethyl ketone
BLS	Basic life support
BMR	Basal metabolic rate
BOCA	Building Officials and Code Administrators
BVM	Bag Valve Mask
C	Centigrade or Celsius
CAMEO	Computer-Aided Management of Emergency Operations
CAS	Chemical Abstract Service
CBC	Complete blood count
CDC	Centers for Disease Control
CFR	Code of Federal Regulations
CGI	Combustible Gas Indicator
CHEMTREC	Chemical Transportation Emergency Center
CISD	Critical Incident Stress Debriefing

CL	Ceiling level
CN	Chloracetephenone
CNS	Central nervous system
CO	Carbon monoxide
COHg	Carboxihemoglobin
COPD	Chronic obstructive pulmonary disease
CPAP	Continuous positive airway pressure
CS	Chlorobenzalmalonitrile
DDT	Dichlorodiphenyltrichloroethane
DECON	Decontamination
DER	Department of Environmental Resources
DMA	Dimethylamine
DMSO	Dimethyl sulfoxide
DO	Designated officer
DOE	Department of Energy
DOT	Department of Transporation
DRI	Direct reading instruments
DUMBELS	**D**iarrhea, **U**rination, **M**iosis, **B**ronchospasm, **E**mesis, **L**acrimation, **S**alivation
EC	Effective concentration
ED	Emergency department
EDC	Emergency dispatch center
ECG	Electrocardiogram
EEL	Emergency exposure limit
EL	Exposure limits
EKG	Electrocardiogram
ERG	Emergency Response Guide
EMS	Emergency medical services
EMT	Emergency Medical Technician
EMT-P	Emergency Medical Technician–Paramedic
EPA	Environmental Protection Agency
ES	Entry supervisor
ET	Effective temperature
F	Fahrenheit
FD	Fire department
FDA	Food and Drug Administration
FEV	Forced expiratory volume
FID	Flame ionization detector
FP	Fire point
FVC	Forced vital capacity
GRS	German Research Society
H	Humature
HBO	Hyperbaric oxygen

HBV	Hepatitis B virus
HEPA	High-efficiency particular air
HF	Hydrofluoric
HIV	Human Immunodeficiency Virus
HM	Hazardous materials
HMIS	Hazardous Materials Information System
HVAC	Heating, ventilation and air-conditioning systems
ICS	Incident command system
IC	Incident commander
IV	Intravenous
IVP	Intravenous push
IDLH	Immediately dangerous to life and health
IMS	Incident management system
IP	Ionization potential
IT	Ignition temperature
IUPAC	International Union of Pure Applied Chemistry
LC	Lethal concentration
LC-low	Lethal concentration–low
LCD	Liquid crystal display
LD	Lethal dose
LEL	Lower explosive limit
LEPC	Local Emergency Planning Committee
LIPE	Life safety, incident stabilization, property conservation, environmental protection
LPG	Liquid petroleum gas
LR	Lacated Ringers
MAC	Maximum allowable concentration
MAK	Maximum allowable concentration
MEK	Methyl ethyl ketone
MAST	Medical Antishock trouser
MetHg	Methemoglobin
mm Hg	Millimeters of mercury
MSDS	Material safety data sheets
MW	Molecular weight
NA	North America
NAERG	North American Emergency Response Guidebook
NFPA	National Fire Protection Association
NRC	National Response Center
NIOSH	National Institute for Occupational Safety and Health Administration
OC	Oleoresin capsicum
O_2Hg	Oxyhemoglobin
ORM	Other regulated material

OSHA	Occupational Safety and Health Administration
PaO_2	Partial Pressure of Oxygen
PCB	Polychlorinated biphenyl
PCC	Poison control centers
PCP	Phenocyclidine
PEL	Permissible exposure limit
PEEP	Positive end expiratory pressure
PFT	Pulmonary function test
PID	Photoionization detector
PIO	Public information officer
PK	Peak value
PPD	Purified protein derivative
PPE	Personal protective equipment
PPM	Parts per million
PRN	*Pro re nata*, Latin for "as needed"
PT	Prothrombin time
RAD	Radiation absorbed dose
REM	Roentgen equivalent man
RD	Respiratory depression
REL	Recommended exposure limit
SaO2	Saturation of oxygen
SAR	Supplied air respirators
SARA	Superfund Amendments and Reauthorization Act (1986)
SERC	State Emergency Response Commission
SCBA	Self-contained breathing apparatus
SG	Specific gravity
SLUD	Salivation, Lacrimation, Urination, Defecation
SLUDGE	**S**alivation, **L**acrimation, **U**rination, **D**efecation, **G**astrointestinal, **E**mesis
SMAC	Sequential multiple analysis chemistry
SOC	Standard of care
SOG	Standard operating guidelines
SOP	Standard operating procedures
STCC	Standard transportation commodity code
STEL	Short-term exposure limit
STP	Standard temperature pressure
TEPP	Tetraethylpyrophosphate
THI	Temperature humidity index
TLC	Toxic concentration, low
TDL	Toxic dose—low
TLV	Theshold limit value
TLV-c	Theshold limit value—ceiling
TLV-s	Theshold limit value—skin

TNT	Trinitrotoluene
TWA	Time weighted average
UN	United Nations
UEL	Upper explosive limit
USCG	Unitied States Coast Guard
VD	Vapor density
VP	Vapor pressure
WBGT	Wet bulb globe temperature